OXFORD

Maths links

Basics Resource Pack

OXFORD
UNIVERSITY PRESS

UNIVERSITY PRESS

Great Clarendon Street, Oxford OX2 6DP

Oxford University Press is a department of the University of Oxford.
It furthers the University's objective of excellence in research, scholarship,
and education by publishing worldwide in

Oxford   New York

Auckland   Cape Town   Dar es Salaam   Hong Kong   Karachi
Kuala Lumpur   Madrid   Melbourne   Mexico City   Nairobi
New Delhi   Shanghai   Taipei   Toronto

With offices in

Argentina   Austria   Brazil   Chile   Czech Republic   France   Greece
Guatemala   Hungary   Italy   Japan   South Korea   Poland   Portugal
Singapore   Switzerland   Thailand   Turkey   Ukraine   Vietnam

Oxford is a registered trade mark of Oxford University Press
in the UK and in certain other countries

British Library Cataloguing in Publication Data

Data available

ISBN 9780199128853

10 9 8 7 6 5 4 3 2 1

Printed in Great Britain by Ashford Colour Press Ltd, Gosport.

Paper used in the production of this book is a natural, recyclable product made
from wood grown in sustainable forests. The manufacturing process conforms
to the environmental regulations of the country of origin.

The images on the front cover are reproduced courtesy of gualtiero boffi/Sthutterstock.com and MaxFX/Shutterstock.com

About this pack

Basics is a unique resource that has been written for teachers of students who are working below NC level 2 in mathematics at the start of Key Stage 3. The content is designed to support the *MathsLinks* series, and the two can be used inclusively alongside each other in the same classroom. The material draws on the P levels, which have been devised for students working below the National Curriculum.

The author is a teacher with over 30 years' experience in supporting students with special needs, and in this pack she has sought to produce motivating material that does not appear patronising or infantile to secondary school students.

Basics comprises a set of activity sheets divided into 20 chapters that complement and support the *MathsLinks* series throughout Key Stage 3. Each chapter contains 10 activity sheets and is split into topic-based units, so that your students have lots of practice at the level they need. The Core activity sheets contained in the printed book are intended to support students working below the National Curriculum.

Accompanying the pack is a free CD-ROM containing PDF files of the activity sheets, that can be displayed, printed and photocopied, or uploaded to a network or VLE. In addition to the Core activity sheets, the CD-ROM contains twenty Extension activity sheets for each chapter, which support progression and consolidate learning through the National Curriculum levels. The activity sheets provide clear differentiation, with the level of difficulty ranging from P6 to NC level 3 so you can choose the level that suits an individual student's needs.

The activity sheets are visual in nature with very few words, thus allowing students to focus on the numerical skills. The emphasis is on repetition with slow and careful progression to reinforce understanding and achievement. The sheets can all be photocopied and written on by the student.

The chapters that make up the pack are:

N1 Adding and subtracting
A1 Sequences
G1 Scales and measurement
N2 Multiplication and division
A2 Symbols and expressions

N4 Fractions and decimals
G3 3D shapes and volume
D3 The probability scale
A4 Balancing and inverse operations
G4 Symmetry and transformations

D1 Collecting and organising data
A3 Equations and substitution
N3 Rounding and factors
G2 2D shapes
D2 Displaying data

N5 Proportion and percentages
D4 Probability and data
A5 Graphs
D5 Statistics and averages
G5 Scale and plans

The activity sheets in *Basics* are arranged by learning objective, ensuring the pack is also suitable for use independently of *MathsLinks*, with groups of students ranging from age 11 to adult.

OXFORD UNIVERSITY PRESS

SINGLE NETWORK MULTIPLE USER ELECTRONIC PRODUCT LICENCE

PLEASE READ THESE TERMS BEFORE USING THE PRODUCT

This Licence sets out the terms on which Oxford University Press ("OUP") agrees to let you (the "Licensee") and your Permitted Users use (i) the copyright work within the media on which this Licence appears, (ii) the associated software embedded within the media, and (iii) any documentation accompanying the media (the "Material"). It is a condition of the Licensee's use of the Material that the Licensee accepts the terms of this Licence. If the Licensee does not accept these terms, the Licensee will be unable to use the Material and should return it to the entity from whom Licensee bought it for a refund of the price.

This licence is a Single Network Multiple User Electronic Product Licence which allows the Material to be networked on a Licensee's single central computer strictly upon the terms and conditions of this Licence.

Definitions of terms are set out either throughout this Licence or at Clause 6.

The content of the Material is © copyright and must not be used, displayed, modified, adapted, distributed, transmitted, transferred or published or otherwise reproduced in any form by any means other than strictly in accordance with the terms of this Licence.

1.	**Licence to use the Material**
1.1	OUP or its licensors own all intellectual property rights in the Material, but grant to the Licensee the non-exclusive, non-transferable licence to allow the Licensee and Licensee's Permitted Users to use the Material upon the terms and conditions of this Licence.
1.2	The Licensee may:

    (a)   load and store the Material on a single central computer (which must not be publicly accessible) (the "Networked Server") which is located at Licensee's single geographical location ("Site") and serves the Licensee's entire internal area computer network ("Network");

    (b)   give Permitted Users whilst located at the Site access to the Material via the Secure Network;

    (c)   where the Network permits, allow Permitted Users whilst located at a third party location access to the Material via the Secured Network;

    (d)   allow Permitted Users who are teachers to load and temporarily store the Material on laptop computer(s) to facilitate the use of the Material solely in connection with bona fide pedagogical purposes;

    (e)   allow Permitted Users to print the Material for personal use, provided that copies are not distributed for profit or commercial use; and

    (f)   allow Permitted Users who are teachers to modify the Material for bona fide pedagogical purposes only.

1.3   Save as set out in this Clause 1, to otherwise copy, alter, republish, post on servers, redistribute, or incorporate the Material in any other work or publication (in whatever format) requires the prior written consent of OUP.

1.4   Recognising the damage to OUP's business which would flow from unauthorised use of the Material, the Licensee will, and procure that its Permitted Users shall, use best endeavours to keep the Material secure during the term of this Licence.

2.   **Warranties**

2.1   OUP warrants that the media on which the Material is supplied will be free from defects on delivery to Licensee.

2.2   Save as provided in Clause 2.1, the Material is provided "as is" and OUP expressly excludes to the maximum extent permitted by law, all other representations, warranties, conditions or other terms, express or implied, including:

    (a)   (save where the Licensee is a consumer), the implied warranties of non-infringement, satisfactory quality, merchantability and fitness for a particular purpose; and

    (b)   that the operation of the Material will be uninterrupted or free from errors.

3.   **Limitations of Liability**

3.1   Save as provided in Clause 3.2, OUP's entire liability in contract, tort, negligence or otherwise for damages or other liability shall be the replacement of the media in which the Material is delivered to the Licensee or return of the price paid for the Material.

3.2   OUP does not seek to limit or exclude liability for death or personal injury arising from OUP's negligence.

4.   **Term and Termination**

4.1   This Licence shall commence on the date that this Licence is accepted by the Licensee and will continue until terminated:

    (a)   by mutual agreement of the Licensee and OUP; or

    (b)   upon the Licensee breaching any of the terms of this Licence.

4.2   Upon termination of this Licence, the Licensee shall cease using the Material and destroy all copies thereof (including stored copies).

5.   **Jurisdiction**

5.1   This Licence will be governed by English Law and the English Courts shall have exclusive jurisdiction.

6.   **Definitions**

6.1   In this Licence unless the context otherwise requires the following definitions apply:

"Permitted Users"   an individual who (a) is employed or engaged under a contract as part of the Licensee's teaching staff (whether on a temporary or permanent basis) or (b) where the Licensee is an academic institution a student of the Licensee (whose normal place of study is at the Site), and (i) whom the Licensee wishes to authorize to have access to the Material; (ii) who can access the Material over the Secure Network; (iii) who has been issued by the Licensee with a valid password; and (iv) who has agreed to be bound by the restrictions on use of the Material contained in this Licence; and

"Secure Network"   secure access to the Network via cable link or a direct secure access medium such as an ISDN link or modem to the Networked Server.

Contents

☑ Core activity sheet
✓ Extension activity sheet

	P6	P7	P8	1c	1b	1a	2c	2b	2a	3
N1 Adding and subtracting										
N1.1 Order and quantities		☑	☑	☑	✓	✓	✓	✓	✓	
N1.2 Mental addition		☑	☑	✓	✓	✓	✓			
N1.3 Mental subtraction		☑	☑	✓	✓	✓	✓			
N1.4 Adding and subtracting	☑	☑	☑	✓	✓	✓	✓	✓	✓	✓
A1 Sequences										
A1.1 Understanding sequences	☑	☑	☑	✓				✓		
A1.2 Numbers and sequences	☑	☑	☑	☑	✓	✓	✓			
A1.3 Sequences and rules	☑	☑	☑	✓	✓	✓	✓	✓	✓	
A1.4 More sequences and rules						✓	✓	✓	✓	✓
A1.5 Creating sequences							✓	✓	✓	✓
G1 Scales and measurement										
G1.1 Time			☑	☑			✓		✓	
G1.2 Units of measurements		☑	☑	✓	✓		✓		✓	✓
G1.3 Reading scales							☑	✓	✓	✓
G1.4 Drawing and measuring lines						☑	✓		✓	✓
G1.5 Compass turns					☑			☑	✓	
G1.6 Angles						☑			☑	✓
G1.7 Parallel and perpendicular lines						✓				✓
G1.8 Solving measuring problems								✓	✓	✓
N2 Multiplication and division										
N2.1 Doubling			☑	☑	✓	✓	✓	✓	✓	
N2.2 Multiply by 10								☑	✓	✓
N2.3 Multiplying and dividing by 10 and 100								☑		
N2.4 Mental and written division			☑	☑			✓	✓		
N2.5 Multiplying and dividing				☑	✓		✓	✓	✓	✓
N2.6 Sharing					☑			☑	☑	
N2.7 More multiplying and dividing							✓		✓	✓
N2.8 Maths Life: Fundraising						✓			✓	✓
A2 Symbols and expressions										
A2.1 Symbols and values		☑	☑	☑	✓		✓	✓		✓
A2.2 Using letters	☑		☑	✓	✓			✓		✓
A2.3 Making expressions		☑	☑	✓		✓	✓	✓	✓	✓
A2.4 Collecting terms		☑	☑	☑	✓	✓	✓	✓	✓	✓
D1 Collecting and organising data										
D1.1 Planning data collection			☑				☑			✓
D1.2 Collecting data		☑	☑					✓	✓	
D1.3 More collecting data			☑	☑	✓	✓	✓	✓	✓	✓
D1.4 Organising data			☑				☑	✓	✓	
D1.5 Data handling		☑	☑				✓	✓	✓	
D1.6 More handling data					✓	✓			✓	
D1.7 The handling data cycle									✓	✓
D1.8 Surveys										✓

	P6	P7	P8	1c	1b	1a	2c	2b	2a	3
A3 Equations and substitution										
A3.1 Substitution			☑	☑	✓	✓		✓	✓	✓
A3.2 Equation problems		☑	☑	☑		☑			✓	✓
A3.3 Equalities			☑	☑	✓		✓			
A3.4 Inequalities			☑		☑				✓	
A3.5 Solving equations						✓		✓		✓
A3.6 More solving equations										✓
A3.7 Two-step problems									✓	✓
A3.8 Solving algebra problems									✓	✓
A3.9 More solving algebra problems										✓
N3 Rounding and factors										
N3.1 Rounding numbers			☑		☑	✓			✓	✓
N3.2 Estimation					☑	✓		✓	✓	
N3.3 Approximation							☑	✓	✓	
N3.4 Factors			☑	☑		✓	✓		✓	✓
N3.5 Multiples and factors			☑	☑	☑	✓		✓		✓
N3.6 Square numbers			☑			✓			✓	✓
N3.7 Prime numbers and factors									✓	✓
G2 2D shapes										
G2.1 Constructing triangles			☑	☑	✓		✓	✓	✓	✓
G2.2 Triangles and area			☑			✓	✓	✓	✓	✓
G2.3 Perimeter and area		☑	☑	☑	☑	✓	✓	✓	✓	✓
G2.4 Compound area			☑	☑	☑	✓	✓		✓	✓
G2.5 More perimeter and area										✓
D2 Displaying data										
D2.1 Drawing pictograms			☑	☑		✓		✓	✓	
D2.2 Displaying data		☑				✓			✓	✓
D2.3 Selecting and drawing charts			☑	☑		✓	✓	✓	✓	✓
D2.4 Tables, graphs and charts	☑	☑	☑	✓		✓			✓	✓
D2.5 Reading charts and diagrams			☑	☑	✓	✓	✓	✓	✓	✓
N4 Fractions and decimals										
N4.1 Fractions			☑		☑	✓	✓	✓	✓	
N4.2 Understanding fractions			☑	✓				✓	✓	✓
N4.3 Fractions and decimals								☑	✓	
N4.4 Ordering decimals			☑	☑	✓			✓		✓
N4.5 Decimal numbers								☑	✓	✓
N4.6 Understanding decimals							☑	✓	✓	✓
N4.7 Decimal multiplication						☑			✓	
N4.8 Decimal addition									☑	✓
N4.9 More decimal addition									✓	
G3 3D shapes and volume										
G3.1 3D shapes		☑	☑	☑	✓	✓	✓	✓	✓	✓
G3.2 Describing 3D shapes				☑		✓	✓	✓	✓	✓
G3.3 Shapes and surface area					☑	☑			✓	✓
G3.4 Surface area of a cuboid						☑	✓		✓	✓
G3.5 Volume				☑		☑			✓	✓
G3.6 Volume of a cuboid						☑			✓	✓

	P6	P7	P8	1c	1b	1a	2c	2b	2a	3
D3 The probability scale										
D3.1 High and low probability		☑	☑	✓	✓					
D3.2 Possibility						☑		✓		
D3.3 The probability scale			☑	☑	☑	✓	✓	✓	✓	✓
D3.4 Understanding probability			☑	☑	✓	✓	✓	✓	✓	✓
D3.5 Equivalent probabilities			☑	☑	✓	✓		✓	✓	✓
D3.6 Describing probabilities									✓	
A4 Balancing and inverse operations										
A4.1 Operation machines				☑	☑	✓		✓	✓	
A4.2 Doing and undoing		☑	☑	✓		✓	✓		✓	
A4.3 Balancing			☑	☑	✓	✓	✓			✓
A4.4 Inverse operations			☑	☑	✓	✓		✓	✓	✓
A4.5 Using inverse				☑	☑	✓		✓	✓	✓
G4 Symmetry and transformations										
G4.1 Symmetry			☑	☑	✓	✓	✓	✓	✓	✓
G4.2 Tessellations				☑	☑	✓		✓	✓	✓
G4.3 Rotation				☑		☑		✓	✓	
G4.4 Transformations						☑	☑	✓	✓	✓
G4.5 Enlargements			☑	☑	✓	✓	✓		✓	✓
N5 Proportion and percentages										
N5.1 Fractions of amounts				☑	☑		✓	✓	✓	✓
N5.2 Ratio			☑	☑	✓	✓		✓	✓	✓
N5.3 Proportion		☑	☑	✓	✓	✓			✓	✓
N5.4 More ratio									☑	✓
N5.5 Percentages						☑			✓	
N5.6 Understanding percentages									☑	✓
N5.7 Percentages of amounts									☑	✓
N5.8 More percentages									✓	✓
D4 Probability and data										
D4.1 Solving data problems	☑	☑	✓	✓	✓	✓	✓	✓	✓	✓
D4.2 Grouped data		☑	☑	✓		✓	✓		✓	✓
D4.3 Probability				☑	☑	✓			✓	✓
D4.4 More probability						☑	☑			✓
D4.5 Outcomes							☑			✓
D4.6 Theoretical and experimental probability							☑			✓
D4.7 Experimental probability										✓
A5 Graphs										
A5.1 Mapping and tables		☑	☑	☑	✓	✓	✓	✓	✓	✓
A5.2 Coordinates					☑	✓		✓		✓
A5.3 Pairs of values								☑	☑	✓
A5.4 Plotting straight – line graphs									☑	✓
A5.5 Straight – line graphs									☑	✓
A5.6 Equation of graphs									☑	✓
A5.7 Parallel lines				☑	✓	✓			✓	
A5.8 More graphs 1										✓
A5.9 More graphs 2										✓
A5.10 More graphs 3										✓
A5.11 More graphs 4										✓

	P6	P7	P8	1c	1b	1a	2c	2b	2a	3
D5 Statistics and averages										
D5.1 The mode		☑	☑			✓	✓	✓	✓	✓
D5.2 Calculating statistics			☑	☑	☑	✓	✓		✓	✓
D5.3 Using statistics							☑	☑	✓	✓
D5.4 Comparing data						☑			✓	✓
D5.5 Making conclusions						☑				✓
D5.6 Discussing findings							☑	✓	✓	
D5.7 The mean and median										✓
D5.8 More calculating statistics						✓			✓	✓
D5.98 Maths Life: The grand opening									✓	✓
G5 Scale and plans										
G5.1 Plans and elevations		☑	☑	☑	✓		✓	✓	✓	✓
G5.2 Scale				☑	☑	✓		✓	✓	
G5.3 Directions				☑	☑		✓		✓	✓
G5.4 Maths Life: A city garden			☑	☑	☑	✓		✓	✓	✓
G5.5 Nets						✓			✓	✓
G5.6 Maths Life: The garden house									✓	✓

Activity sheet objectives

This table includes a learning objective for each activity sheet in the pack and describes how each *Basics* chapter supports the 7A, 8A and 9A books in the *MathsLinks* series.

Unit	Level	Objective
		N1 Adding and subtracting
		Helps students to recall and understand number bonds, develops the skills of counting and teaches the processes of addition and subtraction.
N1.1		**Order and quantities:** develops the students' ability to write and read the value of numbers and relate this skill to familiar objects
	P7	Counting to 5
	P8	Counting to 10
	1c	Ordering numbers to 10 and quantities
N1.2		**Mental addition:** gives students additional practice to aid their understanding and help them recall number bonds to 10 and 20
	P7	Counting to 5
	P8	Matching one to one
N1.3		**Mental subtraction:** develops the students' understanding of subtraction and reinforces their knowledge of number bonds
	P7	Counting to 5
	P8	Practical subtraction with numbers to 5
N1.4		**Adding and subtracting:** consolidates grouping, counting, addition and subtraction
	P6	Group objects
	P7	Count the dots on the dominoes and match the numbers
	P8	Count spanners and match the numbers

This chapter can be used to support
MathsLinks 7A: 1 Integers and decimals, 15 Calculations
MathsLinks 8A:1 Integers, 8 Calculations, 16 Calculating plus
MathsLinks 9A: 7 Calculations

Unit	Level	Objective
		A1 Sequences
		Develops students' ability to observe, identify and continue simple patterns and sequences pictorially and numerically and reinforces the procedures of counting.
A1.1		**Understanding sequences:** assists students in matching and differentiating by size, colour and orientation
	P6	Matching up to 6 simple familiar objects
	P7	Matching groups of real objects
	P8	Matching equal number of dots
A1.2		**Numbers and sequences:** teaches counting and sequencing numbers to in steps to 100
	P6	Counting to 3
	P7	Counting to 8
	P8	Draw objects to 8
	1c	Counting to 10
A1.3		**Sequences and rules:** helps students to match objects and sequences according to the rules
	P6	Match pictures that are the same
	P7	Find the odd one out and match an object from a choice of 5
	P8	Continue a simple sequence based on size

This chapter can be used to support
MathsLinks 7A: 2 Sequences and functions
MathsLinks 8A: 10 Sequences
MathsLinks 9A: 1 Sequences and graphs

G1 Scales and measurement

Helps students to become familiar with time, standard metric units of measurement and compasses. Pupils have the opportunity to look at area, perimeter and angles.

Unit	Level	Objective
G1.1		**Time:** develops students' concepts of time over different periods
	P8	Identifying morning and afternoon and seasons
	1c	Making a judgment on the age of objects—old and new
G1.2		**Units of measurements:** teaches students to use metric measurements
	P7	Circle the smallest - from 6
	P8	Draw a longer/ shorter/ narrower/ taller object
G1.3		**Reading scales:** helps students to interpret sizes given in metric units
	2c	Identify the metric units used to measure volumes and objects
G1.4		**Drawing and measuring lines:** introduces measuring and drawing specific lengths
	1a	Draw and measure lines in centimetres
G1.5		**Compass turns:** introduces students to the major points of the compass
	1b	Moving around a number square by following given directions
	2b	Finding North / South / East / West
G1.6		**Angles:** teaches students to compare the amount of turn in an angle and introduces the terms acute and obtuse
	1a	Larger and smaller angles
	2a	Identifying right angles

This chapter can be used to support
MathsLinks 7A: 3 Measures, 9 Angles
MathsLinks 8A:2 Measures, 6 Angles and shapes
MathsLinks 9A: 3 Geometrical reasoning and construction, 6 Measures

N2 Multiplication and division

Contains exercises in place value and using number lines to aid multiplication and division.

Unit	Level	Objective
N2.1		**Doubling:** helps students to learn the doubles of numbers to 10
	P8	Circling groups of 2
	1c	Turning pictures into sums
N2.2		**Multiply by 10:** uses counting in 10s to reinforce the 10 times tables
	2b	Counting 10p coins
N2.3		**Multiplying and dividing by 10 and 100:** develops skills in adding and subtracting 10 to reinforce the times tables
	2b	Identify the number of 10p coins in a given amount
N2.4		**Mental and written division:** contains exercises in grouping and sharing objects to develop an understanding of division
	P8	Circle groups of objects
	1c	Circle groups of specified quantities
N2.5		**Multiplying and dividing:** helps students to understand odd and even numbers
	1c	Putting objects in simple layout into groups and counting the groups of 2
N2.6		**Sharing:** uses partitioning numbers and table grids to solve division problems
	1b	Sharing money between 2 people
	2b	Finding the half of a number using grids
	2a	Identifying which grids can be shared evenly between 2—the idea of odd and even

This chapter can be used to support
MathsLinks 7A: 1 Integers and decimals , 7 Calculation and measure, 15 Calculations
MathsLinks 8A: 1 Integers, 8 Calculations, 16 Calculating plus
MathsLinks 9A: 7 Calculations

A2 Symbols and expressions		
Assists students in turning real problems into numbers and mathematical expressions and helps students to solve the resulting addition and subtraction sums. This chapter teaches students to count, group and solve simple linear equations.		
Unit	**Level**	**Objective**
A2.1		**Symbols and values:** helps students to count on from a given number
	P7	Counting objects and identifying the numerical symbols to 5
	P8	Counting on from a given number using + and = maths symbols
	1c	Adding on to a given number using numbers and pictures
A2.2		**Using letters:** teaches students to count and simplify equations.
	P6	Circle groups of objects
	P8	Draw the correct number of objects
A2.3		**Making expressions:** teaches students to count and use letters to represent unknown quantities
	P7	Count to 5
	P8	Count to 10
A2.4		**Collecting terms:** teaches students to group objects or number in order to simplify the problems.
	P7	Grouping sets of distinct objects
	P8	Grouping sets of similar objects
	1c	Identifying shapes regardless of size
This chapter can be used to support *MathsLinks 7A*: 6 Operations and symbols, 12 Symbols and expressions, 16 Equations and formulas *MathsLinks 8A*: 5 Expressions and formulae, 13 Algebra *MathsLinks 9A*: 4 Equations, 11 Integer, expression and formulae		

D1 Collecting and organising data		
Teaches students to collect, read, display and interpret data.		
Unit	**Level**	**Objective**
D1.1		**Planning data collection:** assists student with grouping data and identifying sources
	P8	Group like objects
	1a	Count the number of objects
D1.2		**Collecting data:** helps students to identify data and interpret tally charts. Using tally charts will also help to reinforce counting skills.
	P7	Identify sets
	P8	Find the odd one out
D1.3		**More collecting data:** reinforces skills in collecting data and using tally charts
	P8	How many?
	1c	Draw the correct number and colour given number
D1.4		**Organising data:** helps students to organise data using tally charts and bar graphs
	P8	Linking objects to the correct number
	1a	Collecting and organising data from tossing a dice - a dice is required
D1.5		**Data handling:** requires students to count objects and compare data
	P7	Grouping objects
	P8	Counting objects and comparing sizes
This chapter can be used to support *MathsLinks 7A*: 5 Representing data *MathsLinks 8A*: 11 Collecting and representing data *MathsLinks 9A*: 5 Surveys		

A3 Equations and substitution

Teaches students to read, write and use linear equations, understand the relationships between operations and use symbols to compare numbers and equations.

Unit	Level	Objective
A3.1		**Substitution:** introduces substituting objects and numbers in various formats
	P8	Substituting objects for numerals
	1c	Completing a matrix
A3.2		**Equation problems:** teaches students to read and solve simple problems
	P7	Circle the correct number
	P8	Recognise the value of numerals
	1c	Answer simple problems pictorially
	1a	Complete problems
A3.3		**Equalities:** helps students to understand that addition can be done in any order and that a number may be made in several ways, i.e. 7 + 1 and 3 + 5
	P8	Matching quantities to a numerical value
	1c	Matching dominoes of equal value
A3.4		**Inequalities:** helps students to understand size and use the greater than and equals signs
	P8	Identifying the largest and smallest in a group of 3 objects
	1b	Identifying the largest and smallest in a group of 4 shapes

This chapter can be used to support
MathsLinks 7A: 6 Operations and symbols, 12 Symbols and expressions, 16 Equations and formulas
MathsLinks 8A: 5 Expression and formulae, 13 Algebra
MathsLinks 9A: 4 Equations, 11 Integers, expressions and formulae

N3 Rounding and factors

Looks at estimating length, rounding up and down, partitioning, and finding factors and fractions. Students develop an understanding of the relationship between numbers and multiples and the different ways that these can be recorded.

Unit	Level	Objective
N3.1		**Rounding numbers:** rounding numbers up or down to the nearest 10
	P8	More than / less than 5
	1b	Sequencing to 10 and identifying numbers as larger or smaller
N3.2		**Estimation:** encourages students to estimate length
	1b	Estimating which box contains more objects
N3.3		**Approximation:** helps students to understand when to round up or down and teaches them to apply the rules consistently
	2c	Rounding up or down
N3.4		**Factors:** develops understanding of multiplication, division and factors
	P8	Counting the number of objects in a set
	1c	Dividing a set into groups of 2 and counting the groups
N3.5		**Multiples and factors:** helps students to build numbers and number patterns
	P8	Circle groups of objects
	1c	Draw groups of a given number
	1b	Groups
N3.6		**Square numbers:** develops the concept of square numbers
	P8	Putting objects in a square

This chapter can be used to support
MathsLinks 7A: 7 Calculation and measure, 10 Integers and graphs
MathsLinks 8A: 1 Integers, 8 Calculations, 10 Sequences, 16 Calculating plus
MathsLinks 9A: 7 Calculations, 14 Calculation plus

G2 2D shapes

Teaches students to recognise and identify the properties of 2D shapes and introduces how to draw a triangle. This chapter helps students to understand the angles, area and perimeter of triangles and complex shapes.

Unit	Level	Objective
G2.1		**Constructing triangles:** helps students to identify triangles and their properties and shows students how to draw a triangle
	P8	Trace around the triangles
	1c	Find the odd shape
G2.2		**Triangles and area:** teaches students understand the area of a triangle
	P8	Complete the triangles and colour the pattern
G2.3		**Perimeter and area:** helps pupils to understand size, area and perimeter
	P7	Find the largest and smallest from 3 or 4 objects
	P8	Draw round a shape - the perimeter
	1c	Count the number of cubes in a shape - introducing area
	1b	Draw the shape from the description
G2.4		**Compound area:** teaches students to identify rectangles in a complex shape and look at the area and perimeter of the shapes
	P8	Colour patterns made of rectangles
	1c	Colour shapes a given colour
	1b	Copy quilting patterns on to the grid

This chapter can be used to support
MathsLinks 7A: 9 Angles
MathsLinks 8A: 2 Measures, 6 Angles and shapes
MathsLinks 9A: 3 Geometrical reasoning and construction, 6 Measures

D2 Displaying data

Introduce students to the skills of collecting, representing and interpreting data presented on charts in different format and solve problems using knowledge of numbers, calculations, area, graphs and data.

Unit	Level	Objective
D2.1		**Drawing pictograms:** gives students the opportunity to change data into pictograms.
	P8	Drawing the correct number of objects
	1c	Counting the number of objects
D2.2		**Displaying data:** teaches students to create and read graphs from pictograms and frequency tables
	P7	One to one mapping
D2.3		**Selecting and drawing charts:** develops skill in reading data and drawing bar charts
	P8	Draw the correct number to 5
	1c	Draw the correct number of geometric shapes to 10
D2.4		**Tables, graphs and charts:** reinforces students' understanding of reading and building graphs, charts and pictograms
	P6	Circle the groups
	P7	Count objects to 5
	P8	Circle groups of given objects
D2.5		**Reading charts and diagrams:** teaches reading and comparing data on pictograms and bar graphs
	P8	Comparing size and counting blocks on a bar graph
	1c	Read data on a pictogram

This chapter can be used to support
MathsLinks 7A: 5 Representing data, 8 Data and probability
MathsLinks 8A: 11 Collecting and representing data, 15 Analysing and interpreting data
MathsLinks 9A: 5 Surveys, 12 Representing and interpreting statistics

N4 Fractions and decimals

Helps students to recognise fractions, develop knowledge of decimals and express amounts as decimals.

Unit	Level	Objective
N4.1		**Fractions:** helps students to identify, read and write fractions.
	P8	Finding how much is shaded and shading parts of a rectangle
	1b	Identifying the number of parts an object is divided into up to 10
N4.2		**Understanding fractions:** teaches students to match objects, shapes and fractions
	P8	Matching objects by size
N4.3		**Fractions and decimals:** highlights the relationship between decimals and fractions
	2b	Writing money as a decimal fraction (with 10p coins)
N4.4		**Ordering decimals:** helps students to identify and order numbers by size
	P8	Sequence objects by size
	1c	Identify the largest number
N4.5		**Decimal numbers:** teaches students to express tenths as decimals
	2b	Using money to write decimals
N4.6		**Understanding decimals:** reinforces place value and adding 10
	2c	Adding on 10 and putting beads on an abacus
N4.7		**Decimal multiplications:** writing amounts of money using the decimal point
	1a	Count 10p coins
N4.8		**Decimal addition:** helps students to add decimals
	2a	Adding money

This chapter can be used to support
MathsLinks 7A: 4 Fractions and decimals
MathsLinks 8A: 4 Fractions, decimals and percentages
MathsLinks 9A: 2 Fractions, decimals and percentages

G3 3D shapes and volume

Teaches students to recognise and identify the properties of 3D shapes. Pupils look at 3D shapes and their nets. This chapter also introduces students to the volume of cubes and cuboids.

Unit	Level	Objective
G3.1		**3D shapes:** helps students to visualise 2D and 3D shapes
	P7	Trace around shapes
	P8	Identify and colour shapes
	1c	Find the odd one
G3.2		**Describing 3D shapes:** looks at the properties of the different geometric shapes and relate them to everyday shapes
	1c	Match the shape from 4
G3.3		**Shapes and surface area:** teaches students to identify properties of 3 D shapes
	1b	Matching everyday objects to 3D solid geometric shapes
	1a	2D geometric shapes sides and corners
G3.4		**Surface area of a cuboid:** helps students to find the length and width of the different rectangles in a complex shape
	1a	Length, width and number of squares in a rectangle
G3.5		**Volume:** introduces students to the volume of a cuboid in terms of cubes
	1c	How many boxes?
	1a	2D or 3D shape
G3.6		**Volume:** looks at the area and volume of shapes
	1a	Squares in a rectangle and cubes in a cuboid

This chapter can be used to support
MathsLinks 7A: 17Angles and 3-D shapes
MathsLinks 8A: 14 Construction and 3-D shapes
MathsLinks 9A: 13 Three-dimensional shapes

D3 The probability scale

Looks at the probability of events and associated language. Students learn to understand the likelihood of an event occurring and use appropriate language and expressions to describe the possible outcomes.

Unit	Level	Objective
D3.1		**High and low probability:** introduces one to one mapping and simple scale probability
	P7	One to one matching
	P8	Sorting objects according to their purpose
D3.2		**Possibility:** introduces the fact that some things may or may not occur
	1a	Considering true or false
D3.3		**The probability scale:** looks at the likelihood of specific events
	P8	Match the objects
	1c	Match the adult and young
	1b	Yes or No
D3.4		**Understanding probability:** teaches students to identify objects that are the same and different and find the likely outcomes
	P8	Match people
	1c	Find the odd one out
D3.5		**Equivalent probabilities:** helps students to identify equivalent amounts and ratio
	P8	Count to 6
	1c	Colour the correct number

This chapter can be used to support
MathsLinks 7A: 8 Data and probability
MathsLinks 8A: 3 Probability
MathsLinks 9A: 9 Probability

A4 Balancing and inverse operations

Looks at the relationship between numbers, shapes and processes. Students learn to use these relationships to solve problems and make predictions. This chapter also reinforces students' understanding of counting, grouping and solving simple linear equations.

Unit	Level	Objective
A4.1		**Operation machines:** uses addition and subtraction operations to reinforce number bonds and helps students to identify patterns when using the operations.
	1c	Applying the basic operation of next and looking how a calculator forms identifiable numbers
	1b	Applying an operation with simple machines using numbers to 10
A4.2		**Doing and undoing:** takes the concept of opposites a stage further and aims to help students develop an understanding of the relationship between addition and subtraction.
	P7	Developing the idea of opposites, both pictorially and numerically
	P8	Linking pictorial opposites
A4.3		**Balancing:** teaches students to balance each side of an equations
	P8	Find objects that are the same size
	1c	Match objects. numbers on words and dots
A4.4		**Inverse operations:** helps students to understanding of inverse operations
	P8	Matching opposites
	1c	Pictorial addition and subtraction with numbers to 6
A4.5		**Using inverse:** reinforces the relation between operations
	1c	Identify opposites
	1b	Inverse operations

This chapter can be used to support
MathsLinks 7A: 16 Equations and formulas
MathsLinks 8A: 5 Expressions and formulae, 13 Algebra
MathsLinks 9A: 4 Equations, 11 Integers, expressions and formulae

		G4 Symmetry and transformations

Teaches students to draw and recognise reflections and rotations of 2-D shapes. Students learn to identify lines of symmetry and know a shape can come in different sizes. This chapter also helps students to tessellate shapes, identify the faces of a 3-D shape and identify the shape from various angles.

Unit	Level	Objective
G4.1		**Symmetry:** helps students to identify lines of symmetry and reflective and rotational symmetry
	P8	Colour the pattern, remembering the symmetry
	1c	Draw lines of symmetry
G4.2		**Tessellations:** teaches students to tessellate shapes using translation and rotation
	1c	Complete the tessellated patterns
	1b	Use the same shape and tessellate in different ways
G4.3		**Rotation:** looks at rotational symmetry. Students learn to rotate shapes through 90° and complete rotational patterns.
	1c	Rotating an object
	1a	Rotating a shape
G4.4		**Transformations:** reinforces students ability to draw reflections and rotate shapes
	1a	Identify the lines of symmetry
	2c	Translate the shape and draw the reflection at an angle
G4.5		**Enlargements:** helps students to learn that a shape may come in different sizes
	P8	Colour the largest and smallest geometric shape
	1c	Draw a larger geometric shape

This chapter can be used to support
MathsLinks 7A: 13 Transformations and symmetry
MathsLinks 8A: 9 Transformations
MathsLinks 9A: 10 Transformations and scale

		N5 Proportion and percentages

Helps students to relate fractions to division and percentages. Students learn to apply ratios to simple problems and calculate a fraction of a given amount.

Unit	Level	Objective
N5.1		**Fractions of amounts:** helps students to find a fraction of a quantity
	1a	Identifying half of an amount
	1b	Share objects on to plates / bags
N5.2		**Ratio:** teaches students to colour and apply ratios
	P8	Count objects to 5
	1c	Colour by ratio
N5.3		**Proportion:** develops the ability to draw and colour objects according to the ratio
	P7	Put objects in groups of 2 or 3
	P8	Apply instructions - numbers up to 5 - introduce ratios
N5.4		**More ratio:** gives students additional practice with using ratios
	2a	Solve simple ratio problems
N5.5		**Percentages:** introduces students to percentages
	1a	Developing an idea of percentage
N5.6		**Understanding percentages:** helps students to relate percentages and fractions
	2a	Relating percentages, fractions and sharing
N5.7		**Percentages of amounts:** relates percentages to fractions
	2a	Pictorial representation of a percentage

This chapter can be used to support
MathsLinks 7A: 11 Fractions, ratio and proportion
MathsLinks 8A: 4 Fractions, decimals and percentages, 12 Ratio and proportion
MathsLinks 9A: 2 Fractions, decimals and percentages, 14 Calculation plus

		D4 Probability and data

Helps student to solve problems using knowledge of numbers, calculations, area, graphs and data. Pupils also develop an understanding of the language associated with the probability of events and outcomes.

Unit	Level	Objective
D4.1		**Solving data problems:** teaches students to collect and examine data and identify the probability of a given occurrence
	P6	Match the stamps
	P7	Find the matching stamp
D4.2		**Grouped data:** helps students to sort, group and identify data
	P7	Circle objects that are the same
	P8	Count data
D4.3		**Probability:** teaches students to identify possible outcomes
	1c	Identify the correct sentence
	1b	Select the correct word from opposites
D4.4		**More probability:** introduces the language associated with probable outcomes
	1a	Right or wrong?
	2a	Identify correct outcomes
D4.5		**Outcomes:** helps students toidentify the number of possible outcomes
	2a	Listing possible outcomes
D4.6		**Theoretical and experimental probability:** helps students to identify possible outcomes and express them appropriately
	2a	Identify possible outcomes
		This chapter can be used to support *MathsLinks 7A*: 8 Data and probability, 14 Averages *MathsLinks 8A*: 3 Probability, 11 Collecting and representing data *MathsLinks 9A*: 12 Representing and interpreting statistics

		A5 Graphs

Helps students to identify relationships, plot coordinates and apply functions.

Unit	Level	Objective
A5.1		**Mapping and tables:** helps students to identify and apply functions
	P7	Matching numbers
	P8	One to one mapping of objects
	1c	One to one mapping with numerals
A5.2		**Coordinates:** teaches students to read and plot coordinates
	1b	Identify the coordinates on a grid
A5.3		**Pairs of values:** helps students express relationships between numbers as functions
	2b	Complete function tables
	2a	Turn a function table into a graph
A5.4		**Plotting straight-line graphs:** develops students' ability to plot a graph from a table
	2a	Plot a graph from a function table
A5.5		**Straight-line graphs:** students learn to draw a graph from a function table
	2a	Complete function tables using numbers to 100
A5.6		**Equation of graphs:** looks at reading formulas and drawing graphs
	2a	Draw the graph for a simple formula
A5.7		**Parallel lines:** looks at the direction of lines and introduces the idea of parallel lines
	1c	Direction
		This chapter can be used to support *MathsLinks 7A*: 10 Integers and graphs *MathsLinks 8A*: 7 Equations and graphs, 13 Algebra *MathsLinks 9A*: 8 Graphs

		D5 Statistics and averages

Helps pupils to collect, read, display and interpret data. Students also learn to identify the mode and mean.

Unit	Level	Objective
D5.1		**The mode:** looks at reading and comparing data on pictograms and bar graphs and identifying modes
	P7	Draw the correct number to 3
	P8	Count the correct number to 5
D5.2		**Calculating statistics:** teaches students to read bar charts and calculate averages
	P8	Count objects
	1c	Counting on
	1b	Sequencing numbers
D5.3		**Using statistics:** teaches students to find the mode
	2b	Read charts and find the mode
	2a	Find the mode from graphs and charts
D5.4		**Comparing data:** helps students to read data from a table
	1a	Comparing numbers and sequencing according to size
D5.5		**Making conclusions:** students learn to extract and interpret data from a bar chart
	1a	Read and interpret data from a bar chart
D5.6		**Discussing findings:** Using charts and graphs to interpret data and discuss findings
	2c	Reading data from a bar chart

This chapter can be used to support
MathsLinks **7A**: 14 Averages
MathsLinks **8A**: 15 Analysing and interpreting data
MathsLinks **9A**: 12 Representing and interpreting statistics

		G5 Scale and plans

Teaches students to identify the faces of a 3-D shape and identify the shape from various angles. Students look at scale, plans and direction to solve a range of problems.

Unit	Level	Objective
G5.1		**Plans and elevations:** students learn to identify the different surfaces of a 3-D shape and draw the shapes from different elevations
	P7	Match the object to its silhouette
	P8	Match the body part to its print
	1c	Match the object to its overhead view
G5.2		**Scale:** teaches comparing size and using scale
	1c	Draw objects smaller or larger
	1b	Length in blocks
G5.3		**Directions:** teaches students to identify the position of an object including using the 4 points of the compass
	1c	Place objects in given positions i.e. on or in
	1b	Place objects on shelves in the correct position
G5.4		**Maths Life: A city garden:** students compare objects and their numbers and look at maps and plans to find length, perimeter and area
	P8	Compare garden equipment by size
	1c	Count the vegetables in a row to 10
	1b	Count fence panels

This chapter can be used to support
MathsLinks **7A**: 9 Angles, 17 Angles and 3-D shapes
MathsLinks **8A**: 6 Angles and shapes, 14 Construction and 3-D shapes
MathsLinks **9A**: 13 Three-dimensional shapes

Count the number of creatures in each box.

Count the number of tools in each box.

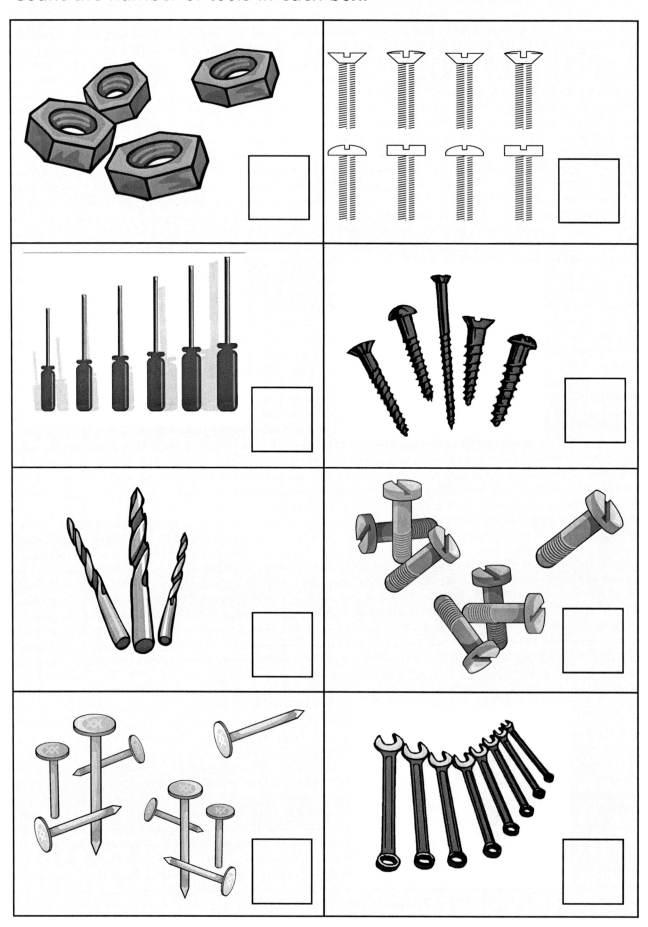

| **N1.1** | **Order and quantities** | **1c** |

Join the bricks in order.

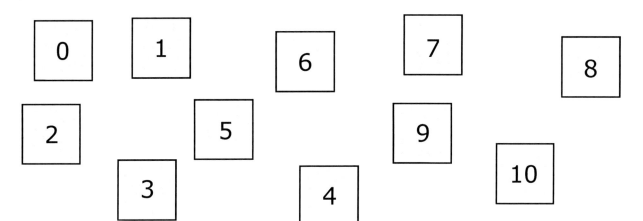

0	1	6	7	8
2	5	9	10
3	4

What are the numbers before and after these numbers?

6	7	8
	3	

	2	
	6	

Draw the correct number of triangles △ in each row.

5	
8	
3	
4	
7	
9	

How many?

How many?

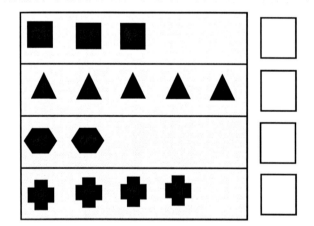

Colour the correct number of squares.

4

3

5

Put 1 pocket on each apron.

Put 2 buttons on each jacket.

Put 3 stripes on each jumper.

Put a plant in the pot.

Put a straw in each glass.

Draw a knife for each fork.

Draw a key for each lock.

| **N1.3** | **Mental subtraction** | **P7** |

How many?

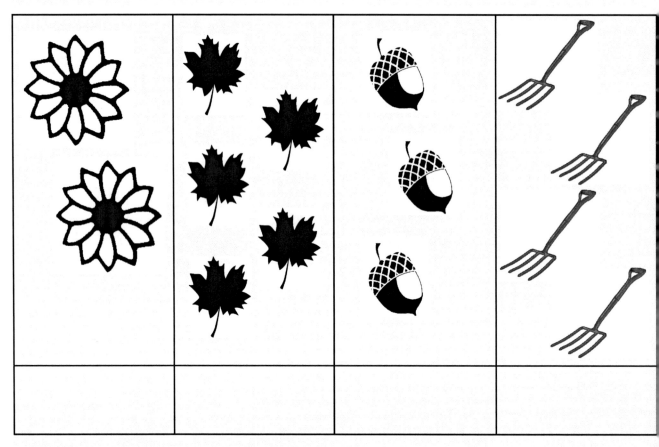

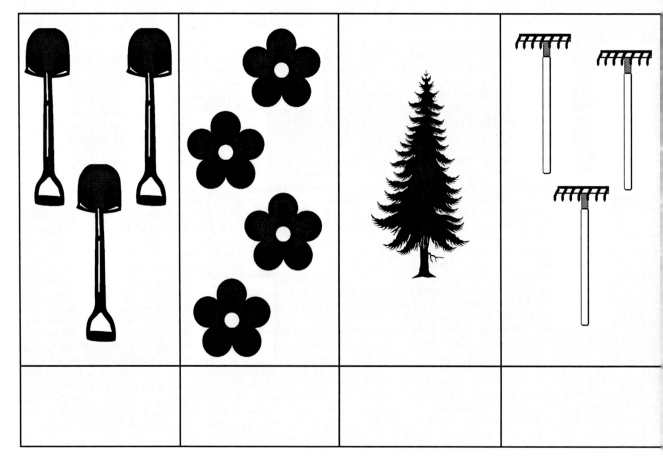

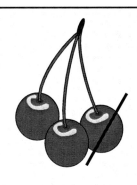 How many? __3__

Cross out 1.

How many left? __2__

 How many? _____

Cross out 1.

How many left? _____

 How many? _____

Cross out 1.

How many left? _____

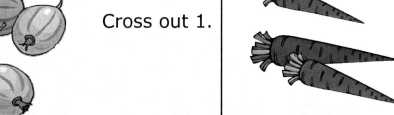 How many? _____

Cross out 1.

How many left? _____

 How many? _____

Cross out 1.

 How many left? _____

How many? _____

Cross out 1.

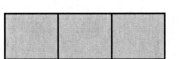 How many left? _____

How many? _____

Cross out 1.

How many left? _____

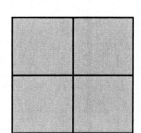 How many? _____

Cross out 1.

How many left? _____

Circle the objects that are the same.

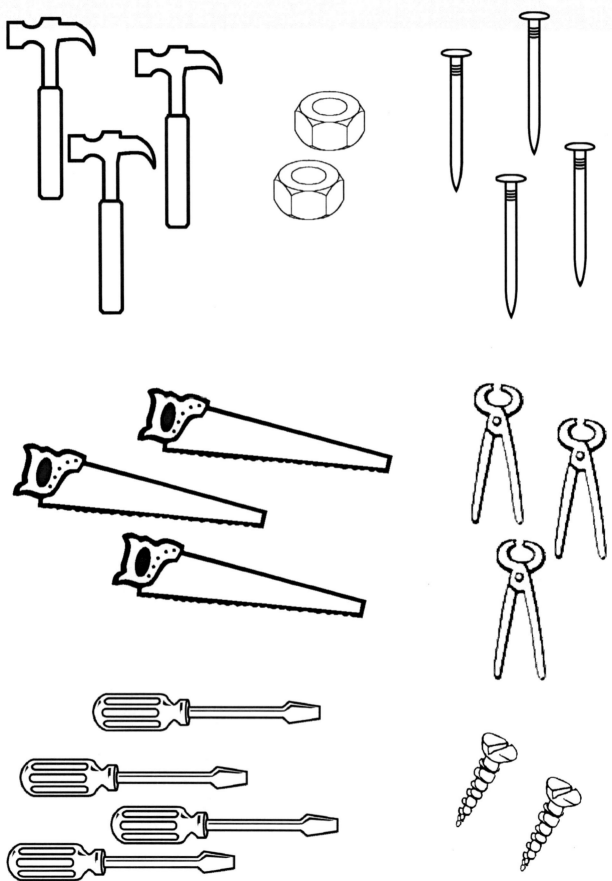

| N1.4 | Adding and subtracting | P7 |

Count each set of tools and link to the correct number.

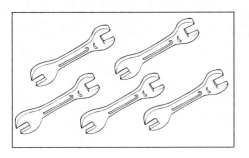

1

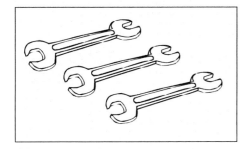

2

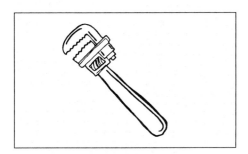

3

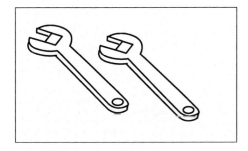

4

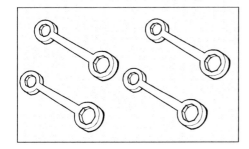

5

Count dots on each domino and link to the correct number.

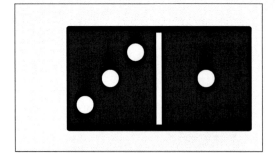

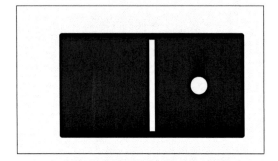

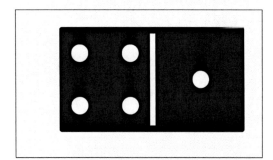

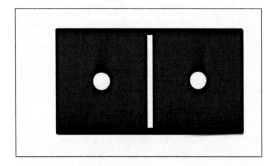

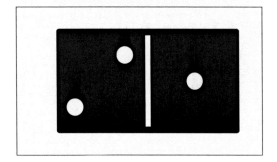

1

2

3

4

5

Link the matching animals.

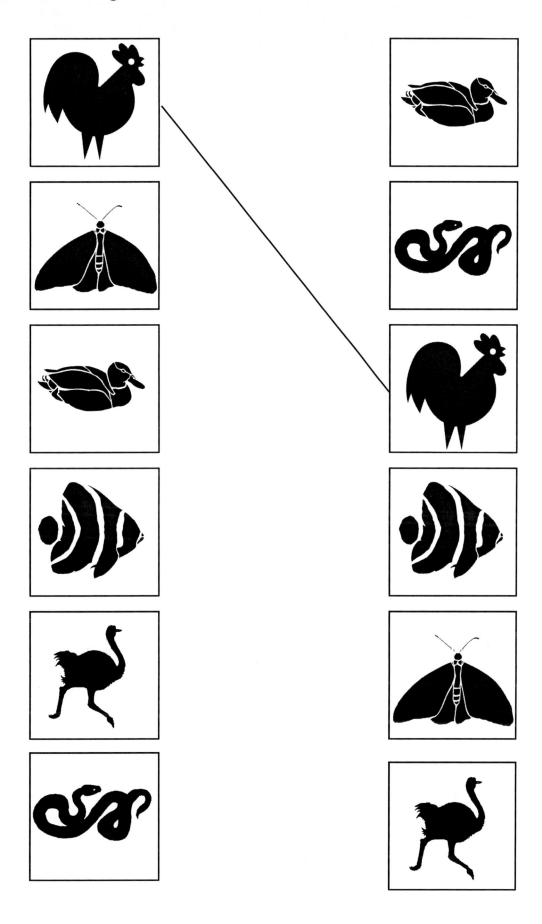

Link the matching spanners.

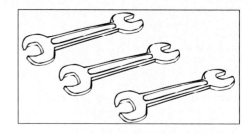

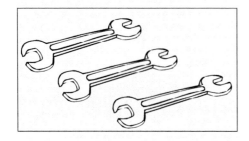

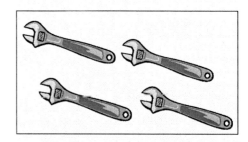

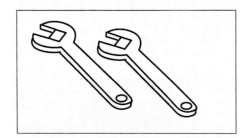

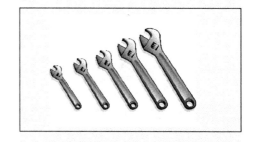

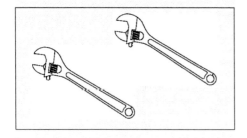

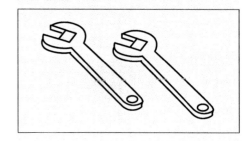

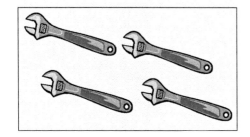

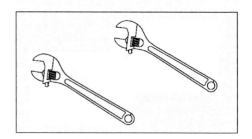

Link the rectangles with the same number of dots.

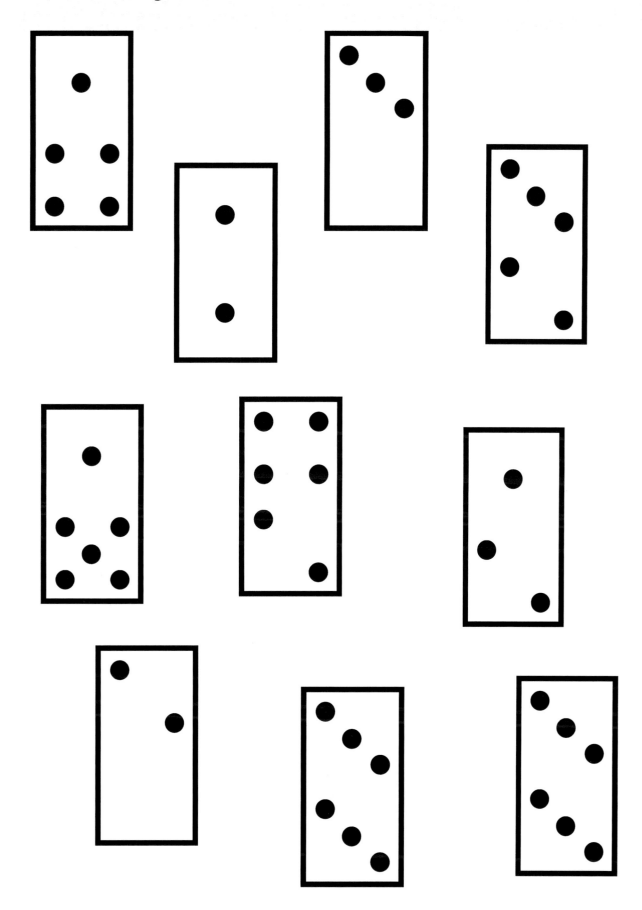

Colour
1 lion

Colour
3 monkeys

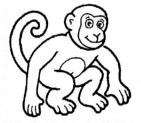  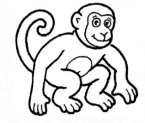

Colour
2 kangaroos

Colour
2 snakes

Colour
3 ladybirds

Copy the numbers.

1 __   2 __   3 __

How many bees in each box?

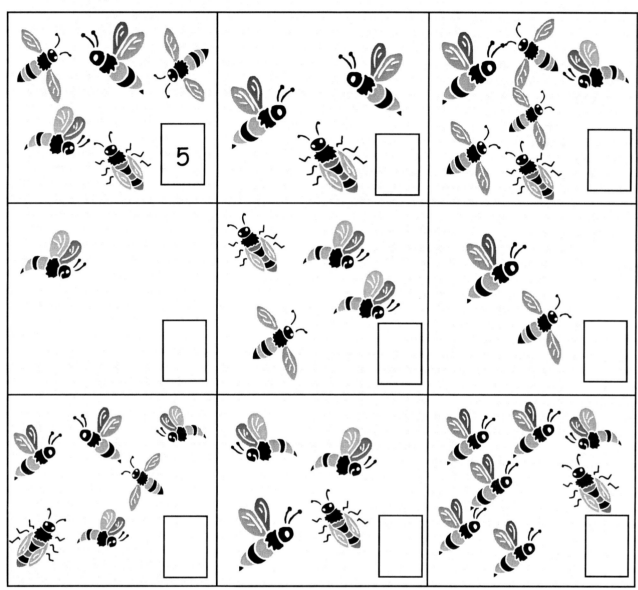

Complete the sequences.

| 1 | | 3 | 4 | 5 | | 7 | 8 |

| | 2 | | 4 | | 6 | | 8 |

| 1 | | | 4 | | | 7 | 8 |

A1.2	Numbers and sequences	P8

Draw 5 □	□ □ □ □ □
Draw 3 △	
Draw 2 ☆	
Draw 7 ○	
Draw 4 ◇	
Draw 8 ▯	
Draw 1 ⬡	
Draw 6 ◺	

| A1.2 | **Numbers and sequences** | 1c |

How many children in each box?

Complete the number sequences.

| 1 | | 3 | | 5 | | 7 | | 9 | |

| | 2 | 3 | | 5 | 6 | | 8 | | |

| | 2 | | | | 6 | | | | 10 |

Matching
Link the matching bolts.

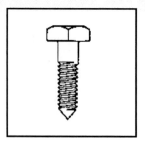

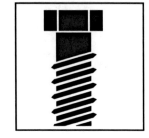

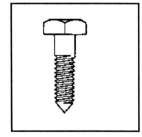

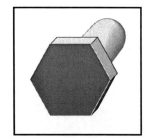

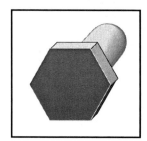

| **A1.3** | **Sequences and rules** | **P7** |

Circle the odd one out in each row.

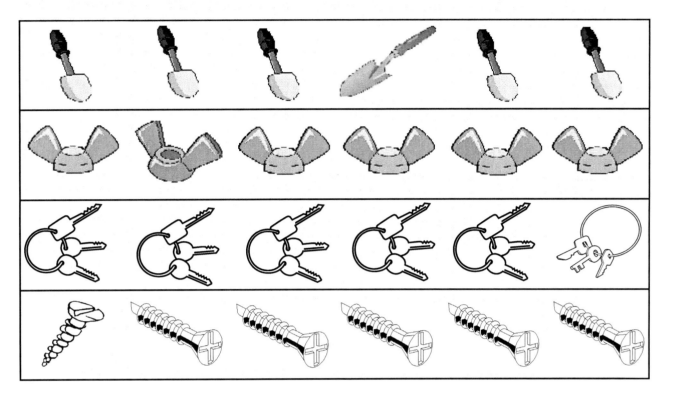

Circle the one that matches the object in the box.

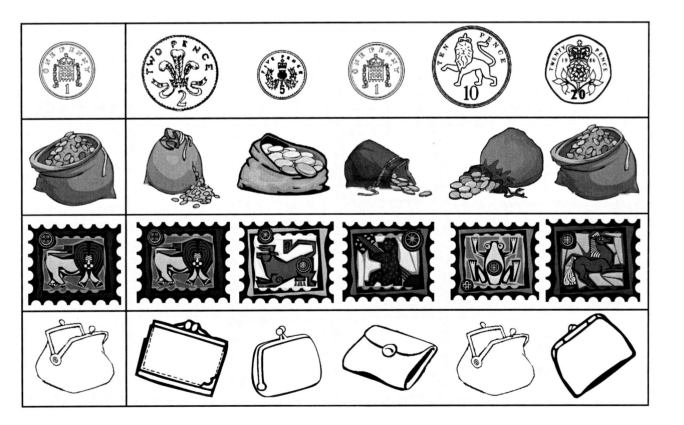

Draw the next picture in each row.

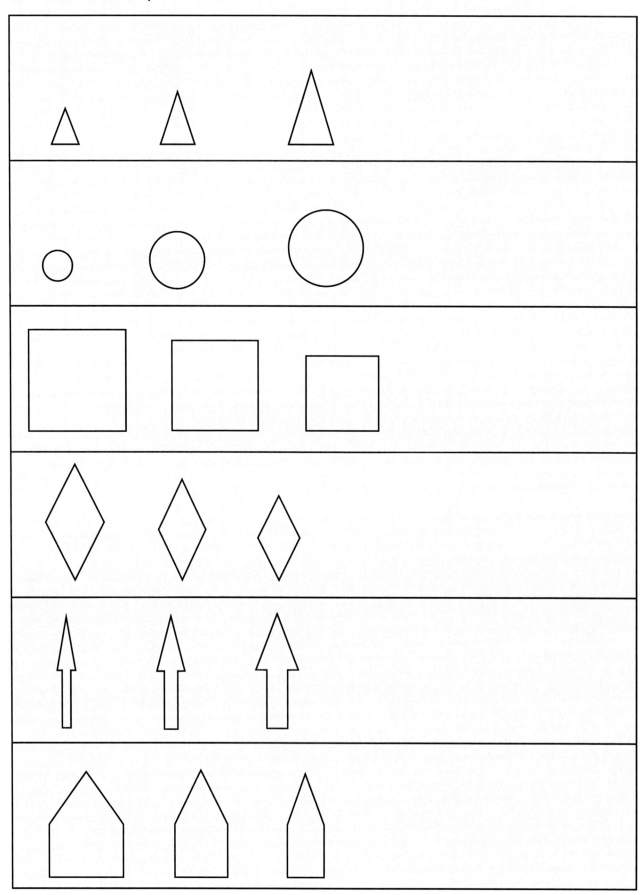

Draw 2 things you do each morning and
2 things you do in the afternoon.

Morning	
Afternoon	

When would you see these? Circle the correct word.

winter summer	winter summer	winter summer
winter summer	winter summer	winter summer

G1.1 | **Time** | **1c**

Link the objects to the correct word.

G1.2 | **Units of measurement** | **P7**

Circle the **smallest** shape in each row.

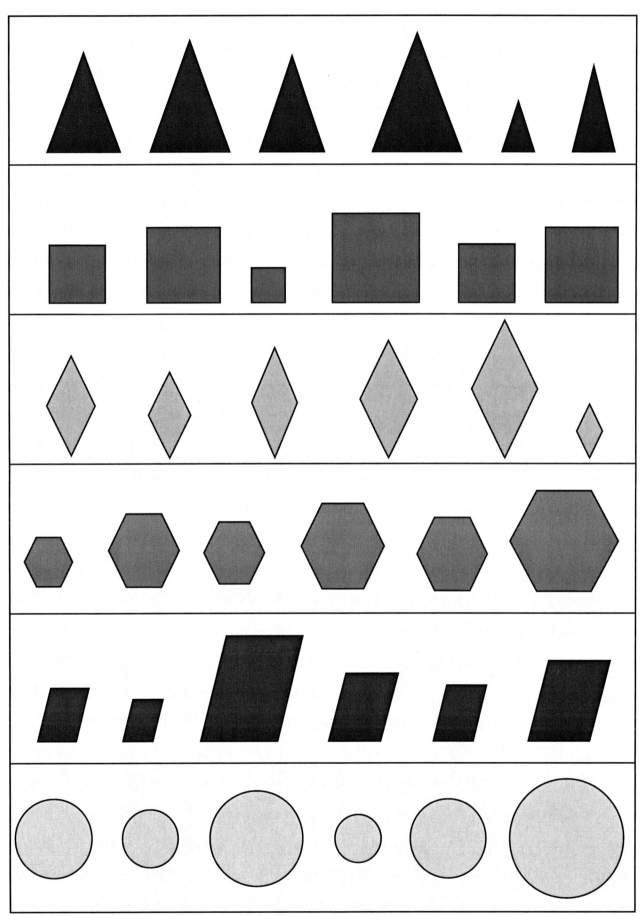

| G1.2 | Units of measurement | P8 |

Draw a **longer** caterpillar. 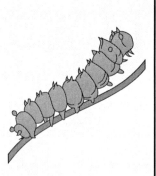	Draw a **smaller** snail. 
Draw a **shorter** fish. 	Draw a **taller** penguin. 
Draw a **wider** feather. 	Draw a **larger** pig. 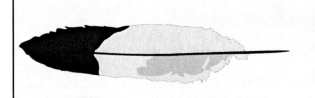
Draw a **narrower** cage. 	Draw a **shorter** giraffe. 

Circle the correct phrase.

	a litre of milk a metre of milk a gram of milk		a litre of apples a metre of apples a kilogram of apples
	10 metres of water 10 millilitres of water 10 centimetres of water		a metre of fabric a litre of fabric a gram of fabric
	20 litres of knitting 20 centimetres of knitting 20 grams of knitting		250 metre bottle 250 gram bottle 250 millilitre bottle
	243 grams of cheese 243 litres of cheese 243 centimetres of cheese		175 millilitres 175 millimetres 175 grams
	49 kilograms 49 kilometres 49 litres		73 kilograms 73 kilometres 73 litres

G1.4 | **Drawing and measuring lines** | **1a**

Use a ruler and draw a line between the dots. How long is the line?

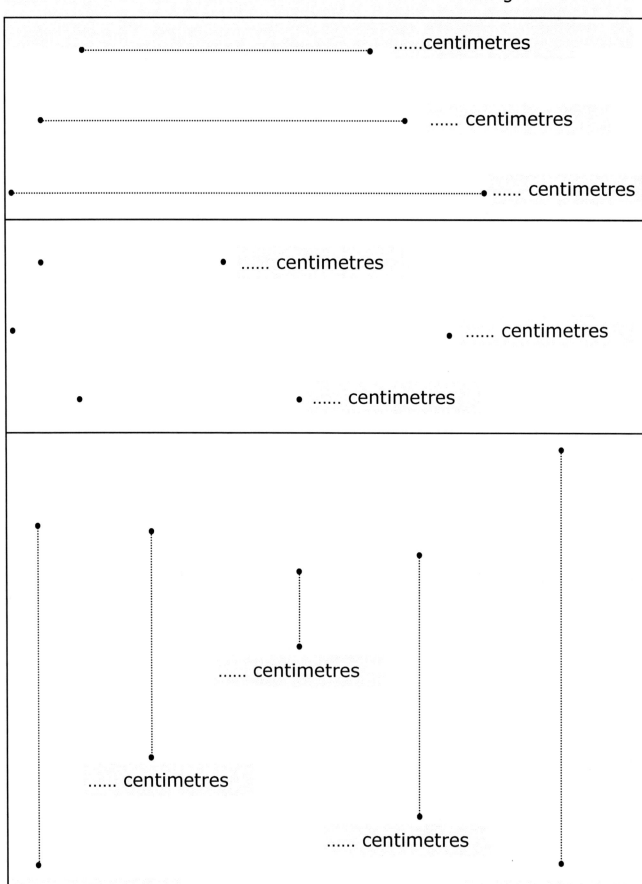

•————————• ......centimetres

•————————• ...... centimetres

•————————• ...... centimetres

• • ...... centimetres

• • ...... centimetres

• • ...... centimetres

...... centimetres

...... centimetres

...... centimetres

...... centimetres

...... centimetres

Follow the trail.

21	22	23	24	25
20	19	18	17	16
11	12	13	14	15
10	9	8	7	6
1	2	3	4	5

Start at 1   ↑3   →4   ↓2   You are on square ....

Start at 21   ↓2   →3   ↓1   ←1   You are on square ....

Start at 5   ←2   ↑3   ←1   ↓2   You are on square ....

Start at 1: up 4, right 3, down 1, left 1. You are on square ....

Start at 13: up 2, left 1, down 3, left 1. You are on square ....

How do I get from 3 to 23?

How would you move from 1 to 14?

How would you move from 25 to 2?

In which direction do the roads go - North-South or East-West?
Tick the right box.

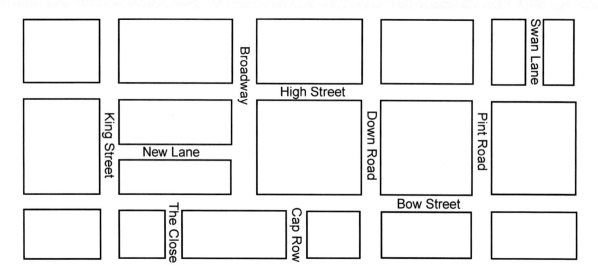

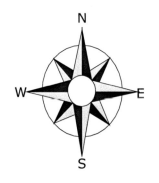

	North-South	East-West
High Street		
King Street		
Swan Lane		
Cap Row		
New Lane		
Tin Close		
Bow Street		

If you travel North up Tin Close and turn East, which street

are you on? ....

If you travel East along New Lane and then turn South, which

road will you be on? ....

If you travel South along Swan Lane and turn West, which

street will you be on? ....

Cross out the smaller angle in each box.

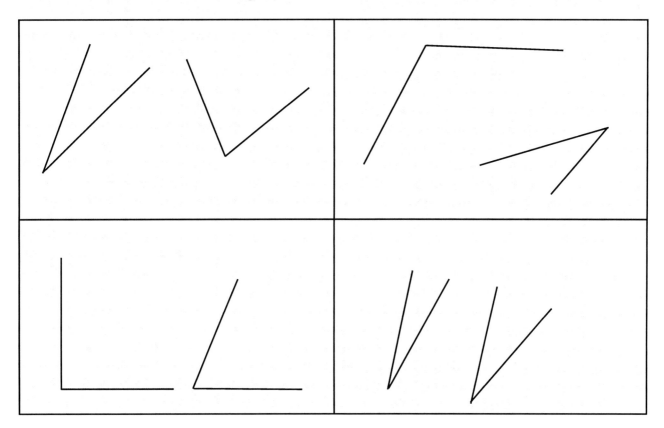

Draw a larger angle in each box.

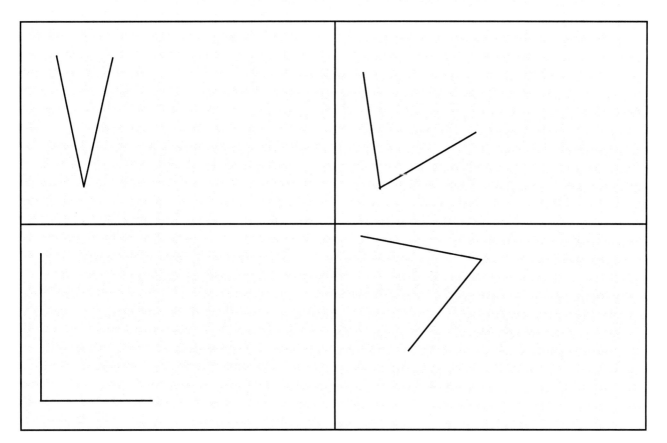

G1.6	Angles	2a

A right angle is a $\frac{1}{4}$ turn, or 90°.
You mark a right angle with a small square.
Other angles are marked with an arc.

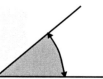

Mark the angles in these shapes.

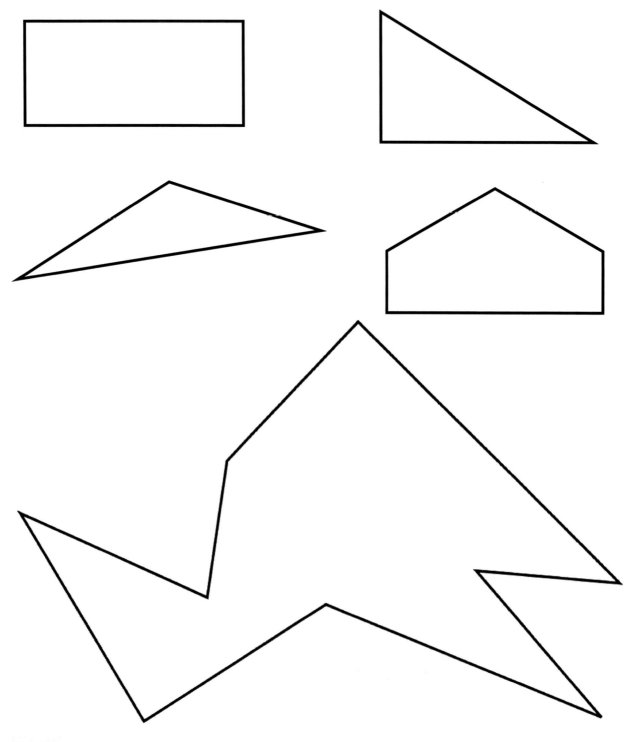

Put circles around sets of 2 in each box.
Count how many sets of 2 in each box.

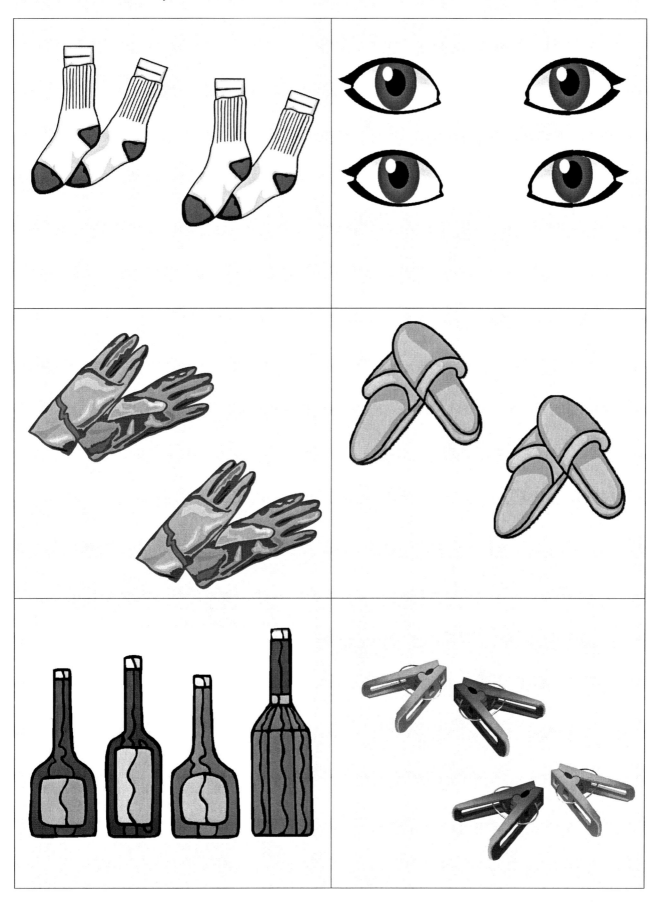

| N2.1 | Doubling | 1c |

Write these pictures as sums.

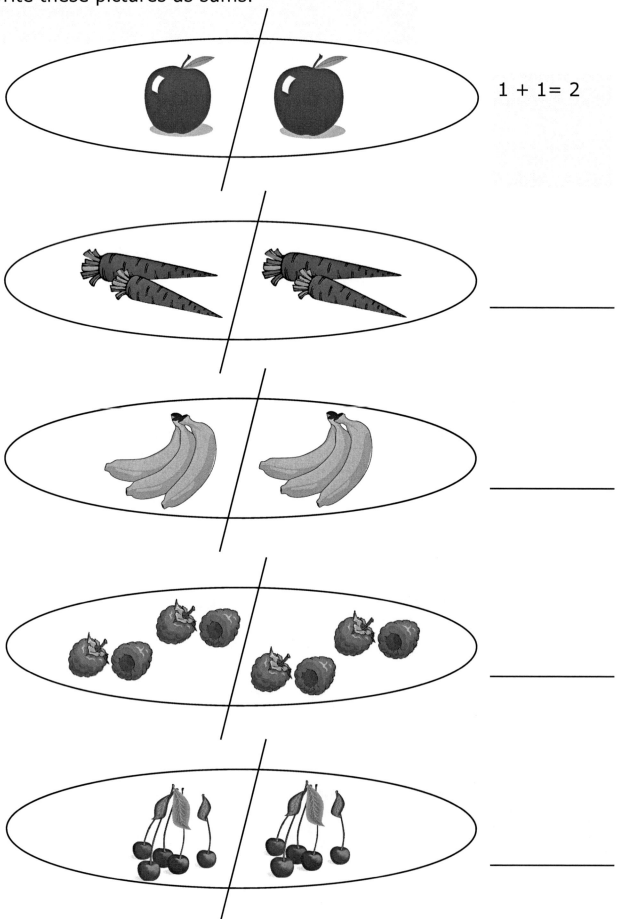

$1 + 1 = 2$

_____

_____

_____

_____

| N2.2 | **Multiply by 10** | 2b |

How much is in each box? Count in 10s.

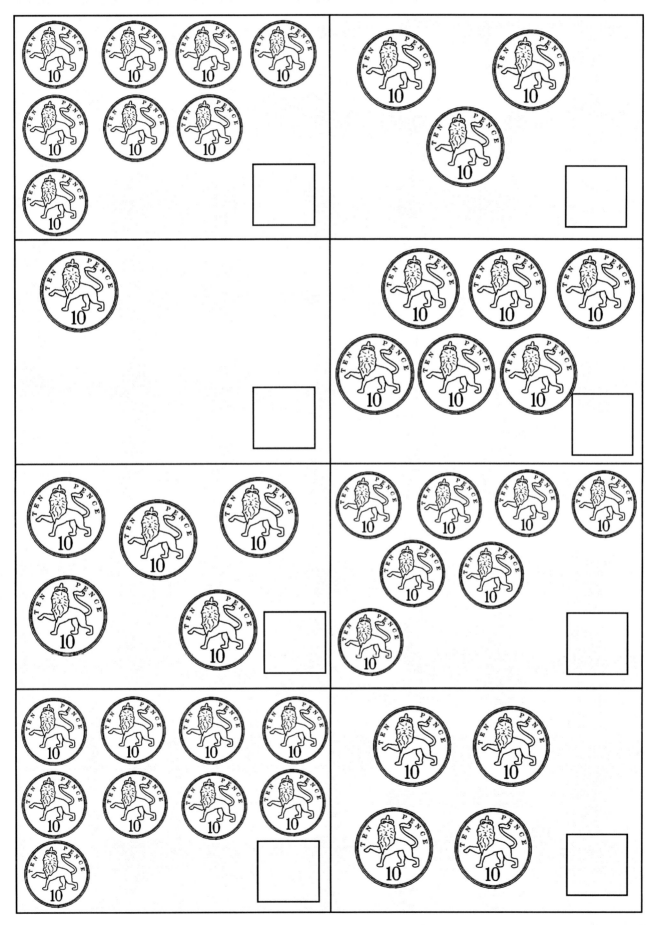

| **N2.3** | **Multiplying and dividing by 10 and 100** | **2b** |

Draw the same amount in 10p coins.

Circle the groups.

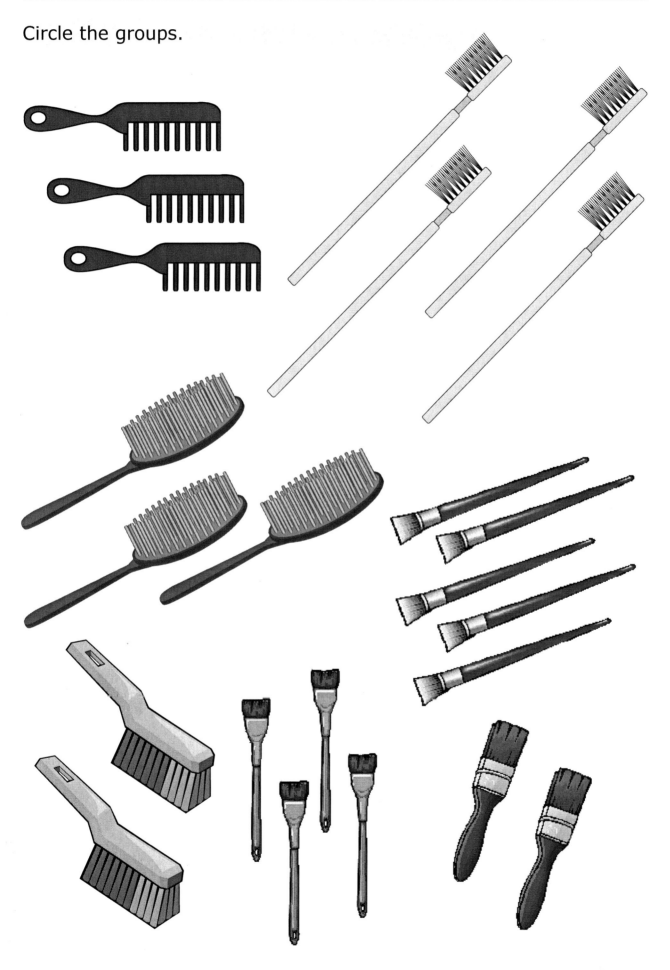

N2.4 | **Mental and written division** | **1c**

Divide into groups.

Groups of 2	Groups of 3	Groups of 4

Groups of 5	Groups of 6	Groups of 7

| **N2.5** | **Multiplying and dividing** | **1c** |

How many groups of two in each box?

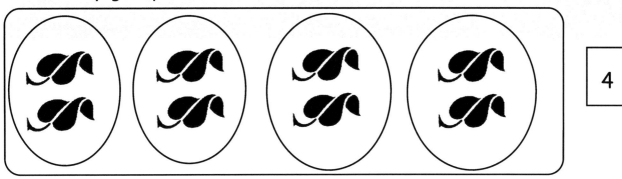

4

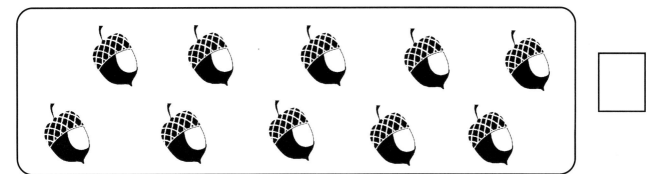

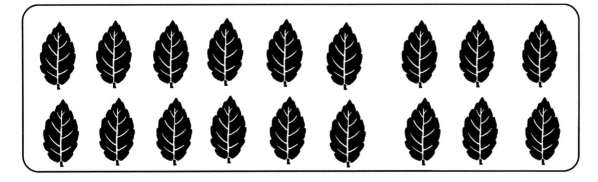

Share the pennies between the two children.

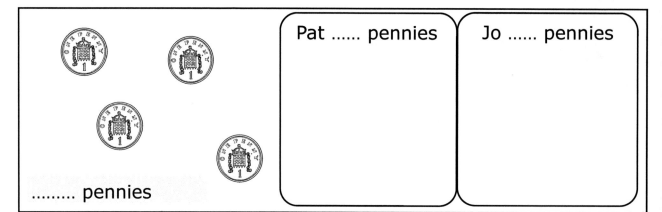

.... pennies

Pat ...... pennies

Jo ...... pennies

.... pennies

Sue ...... pennies

Ken ...... pennies

.... pennies

Tom ...... pennies

Lyn ...... pennies

.... pennies

Ian ...... pennies

Pam ...... pennies

N2.6   **Sharing**   **2b**

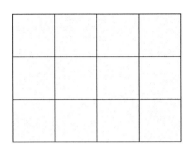

Colour half of the boxes red.

Total number of boxes ....

Number of red boxes ....

Half of 12 =

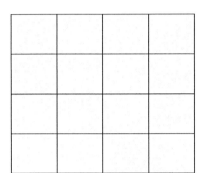

Colour half of the boxes blue.

Total number of boxes ....

Number of blue boxes ....

Half of 16 =

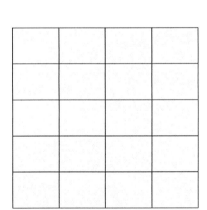

Colour half of the boxes green.

Total number of boxes ....

Number of green boxes ....

Half of 20 =

Continue the sequence of bricks.

  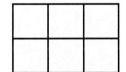

Can you share these bricks evenly between 2 people?
Tick the correct box.

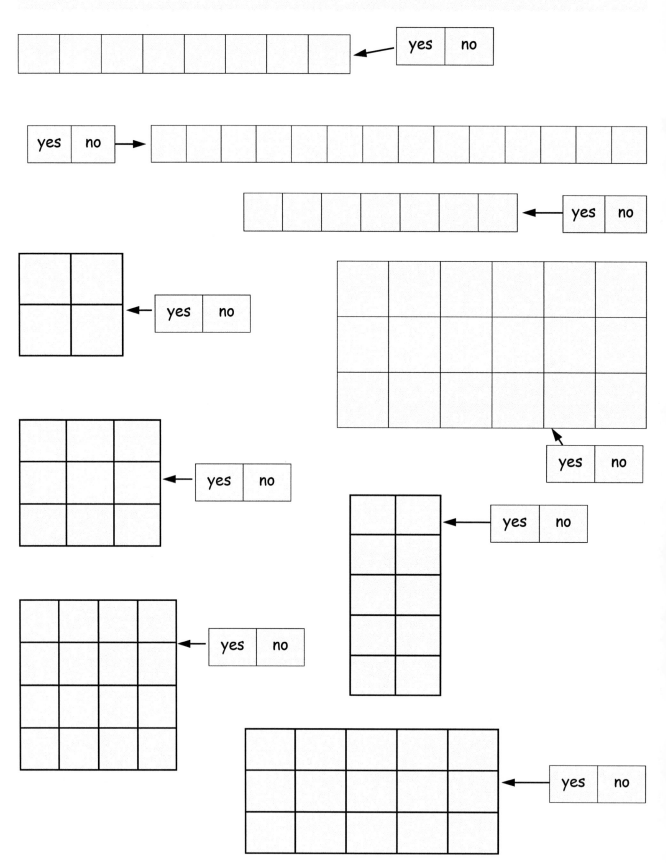

A2.1	Symbols and values	P7

Colour the correct number in each row.

3	
5	
1	
2	
4	
3	
4	

A2.1	**Symbols and values**	**P8**

Add on the objects in each row.

4 + 		= 5
2 +		=
1 +		=
3 +		=
2 +		=
1 +		=
3 +		=
1 +		=
2 +		=

| A2.1 | Symbols and values | 1c |

Are there enough balls in each row? Add more if needed.

8	
4	
6	
3	
5	
7	
9	

Circle the objects that are the **same**.

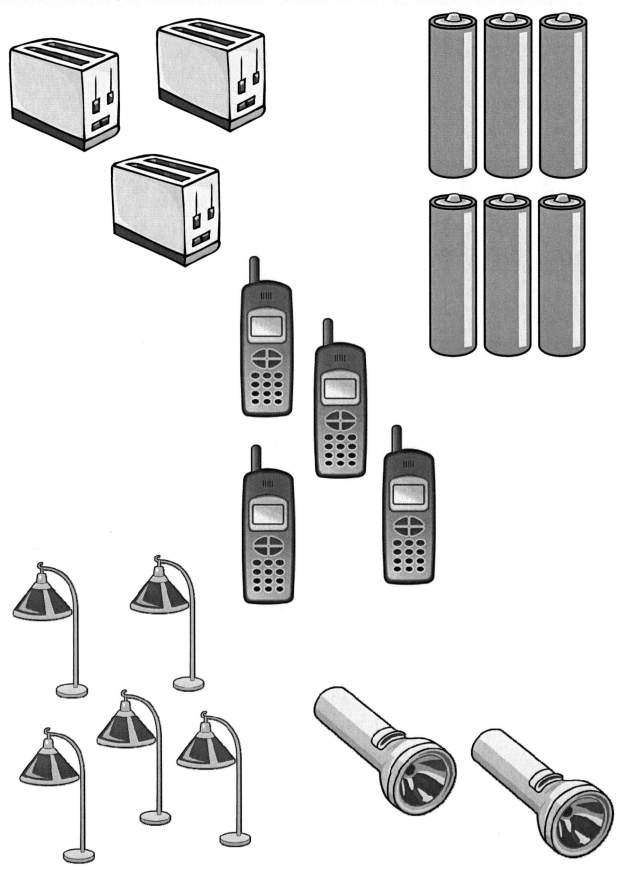

| **A2.2** | **Using letters** | **P8** |

Draw the correct number.

Draw 3 ○	Draw 6 □
Draw 9 △	Draw 7 ▭
Draw 5 ◇	Draw 8 ⬠
Draw 4 ☆	Draw 2 ◺

How many in each group?

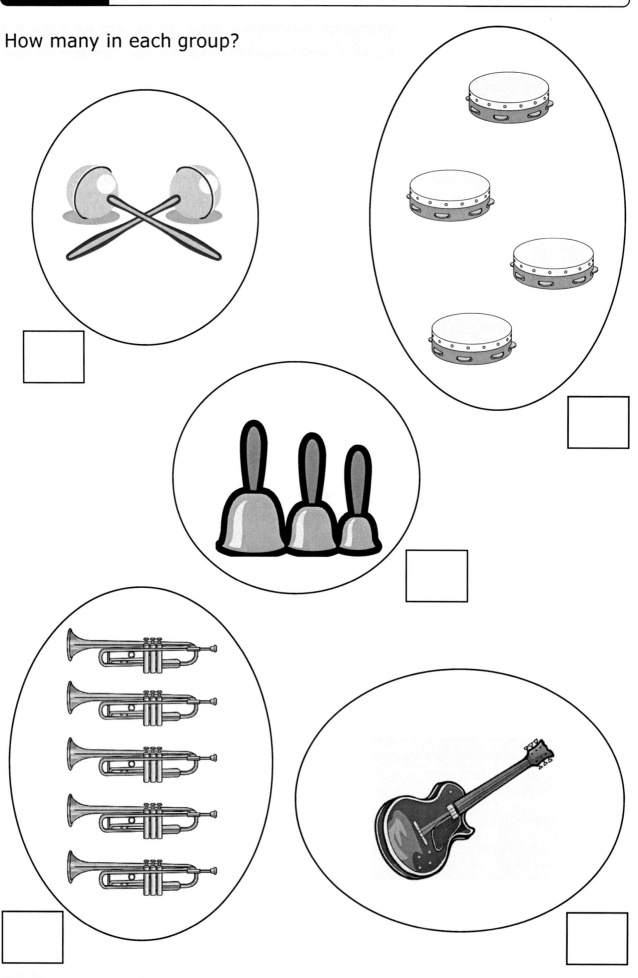

How many in each group?

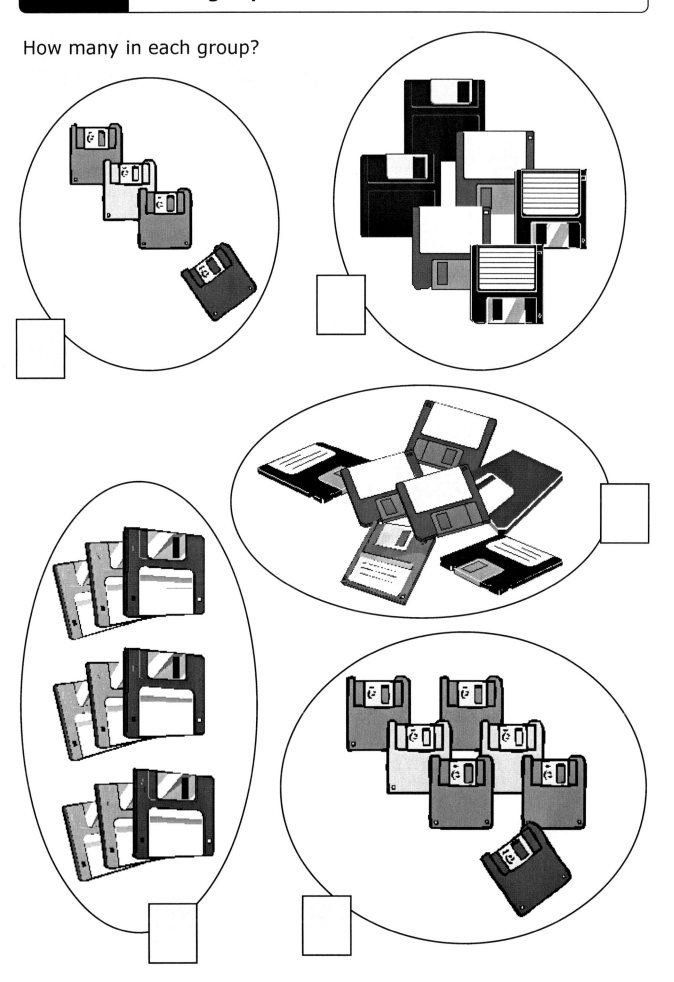

Circle the objects that are the same.

Circle the objects that are the same.

A2.4   **Collecting terms**   **1c**

Colour the circles red ◯.   Colour the squares green ▢.
Colour the triangles blue △.

© Oxford University Press 2011: this may be reproduced for class use solely for the purchaser's institute

Group the objects .

Counting
Count how many of each item you can see on the page.

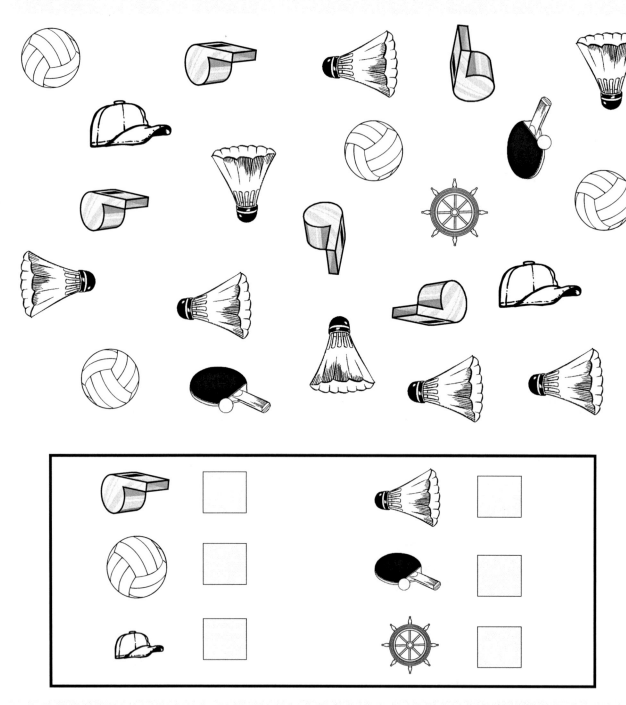

Put the six items in order, starting with the smallest number.

Circle the groups.

Find the odd one out in each row.

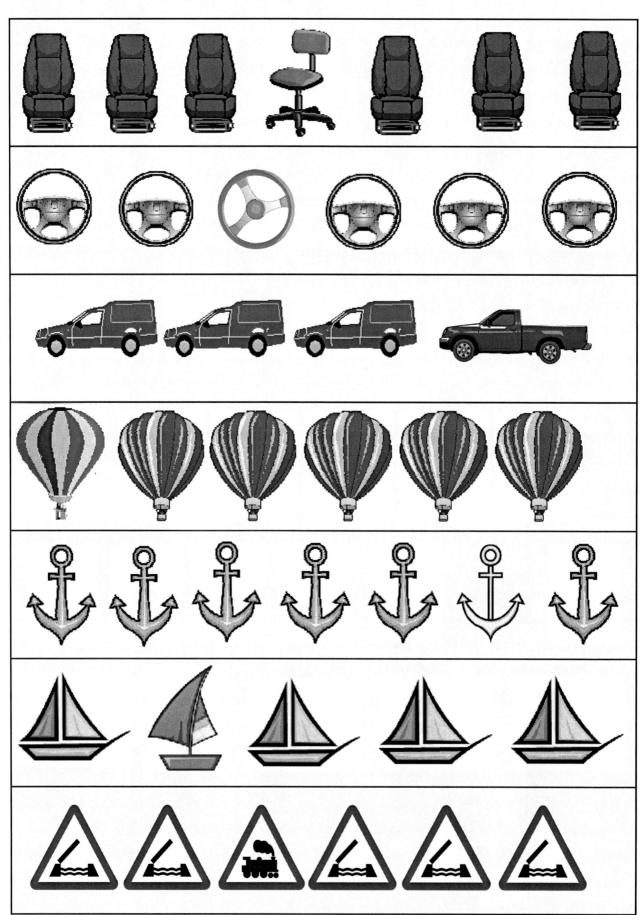

| **D1.3** | **More collecting data** | **P8** |

How many in each row?

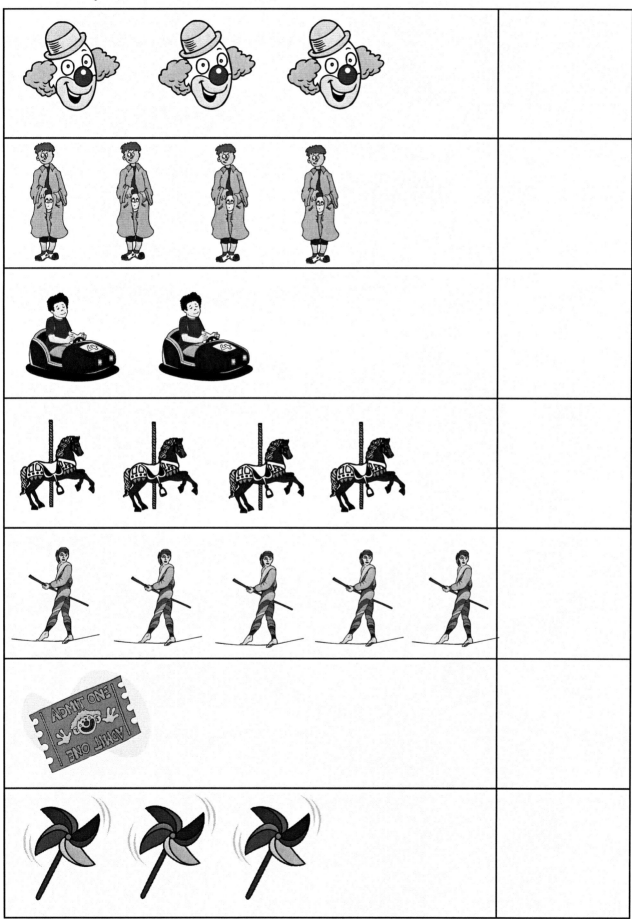

| D1.3 | More collecting data | 1c |

Draw 5	Colour 3 green	Draw 4	Colour 1 orange
Draw 3	Colour 2 blue	Draw 2	Colour 1 brown
Draw 5	Colour 4 yellow	Draw 4	Colour 2 red

| D1.4 | Organising data | P8 |

Link the button boxes to the correct number.

| 1 |
| 2 |
| 3 |
| 4 |
| 5 |
| 6 |
| 7 |
| 8 |
| 9 |
| 10 |

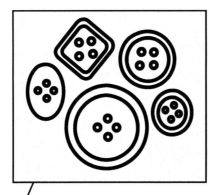

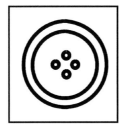

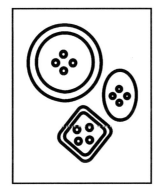

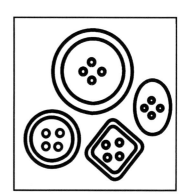

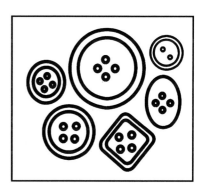

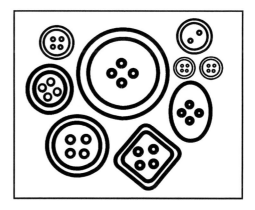

Put a ring around the objects that are the same.

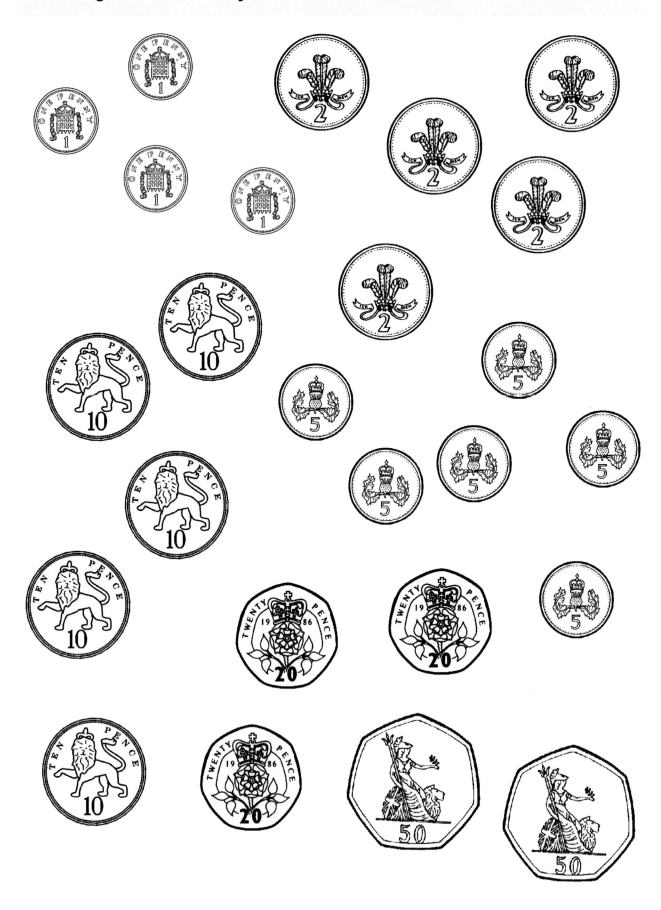

How many bricks in each tower?

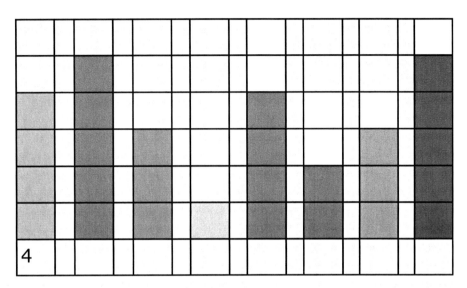

Colour the **tallest** shape red and the **shortest** shape green in each box.

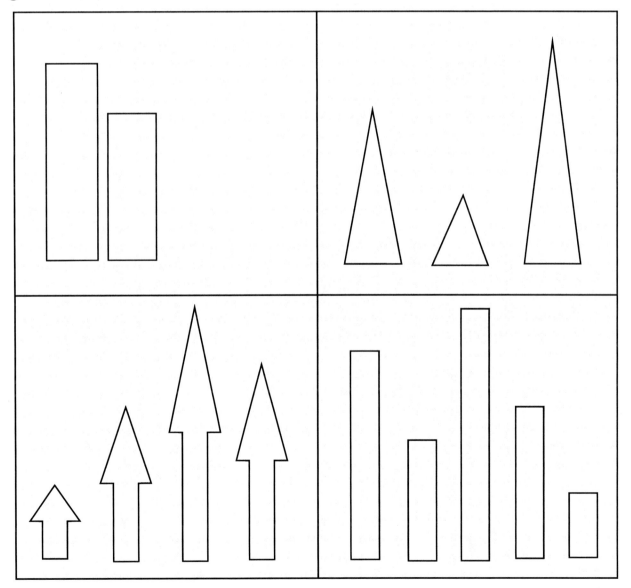

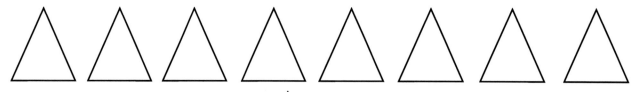

Put a tick ✓ in each triangle. △
How many ticks did you draw? ....

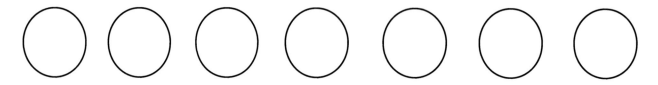

Put a triangle △ in each circle.
How many triangles did you draw? ....

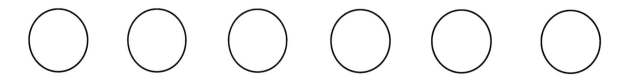

Put a square ☐ in each circle.
How many squares did you draw? ....

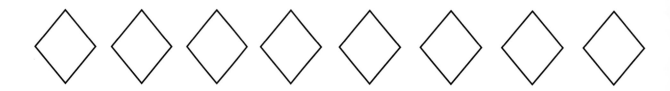

Put a circle ◯ in each diamond.
How many circles did you draw? ....

Draw the right number.

5	
3	
4	
1	
6	
2	
7	

Colour the buttons and give them
the correct number of holes.

Colour	Number of holes	
	2	4
red	◯	◯
green	◯	◯
blue	◯	◯
yellow	◯	◯
orange	◯	◯
brown	◯	◯

How many aliens?

Colour the correct number of objects.

4	
3	
5	
2	
6	
7	

Draw the correct number of shapes.

Draw 2 + 1 triangles	Draw 3 + 2 circles
Draw 4 - 1 squares	Draw 5 - 1 diamonds
Draw 4 + 2 triangles	Draw 6 - 2 circles

◯	Draw 6 circles. Colour 4 green.	△	Draw 5 triangles. Colour 3 blue.
☐	Draw 7 squares. Colour 2 yellow.	◇	Draw 4 diamonds. Colour 1 red.

Use the number line to help fill the table.

1	2	3	4	5	6	7	8	9	10

2 less	1 less	Number	1 more	2 more
		3		
		4		
		5		
		6		
		7		

Match the pictures with the numbers.

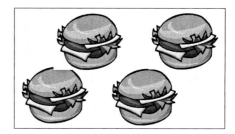

| 1 |
| 2 |
| 3 |
| 4 |
| 5 |
| 6 |
| 7 |
| 8 |
| 9 |

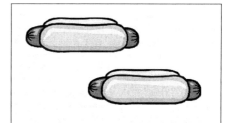

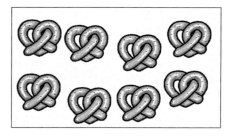

Match the dominoes with the same number of dots.

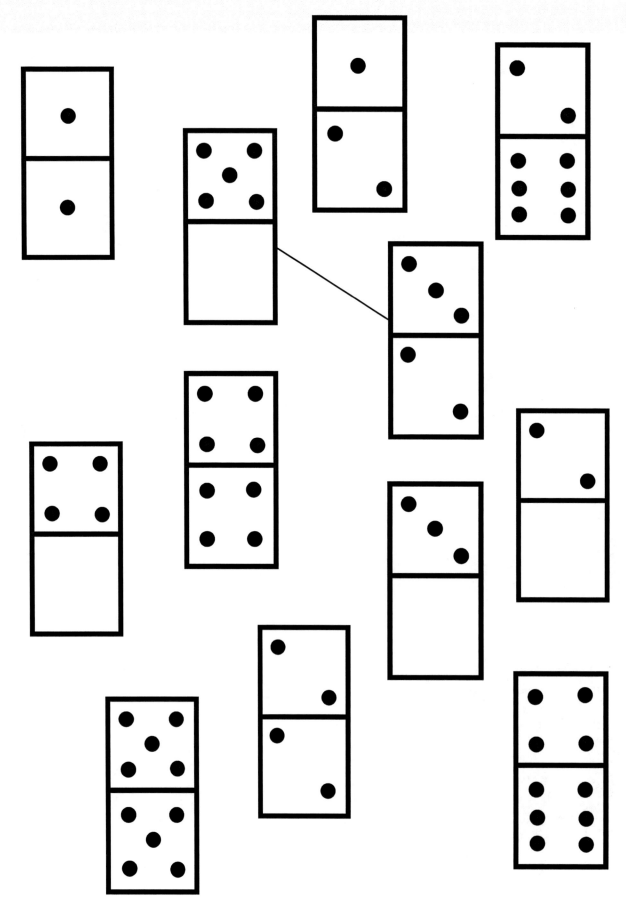

Cross out the smallest and colour the largest in each box.

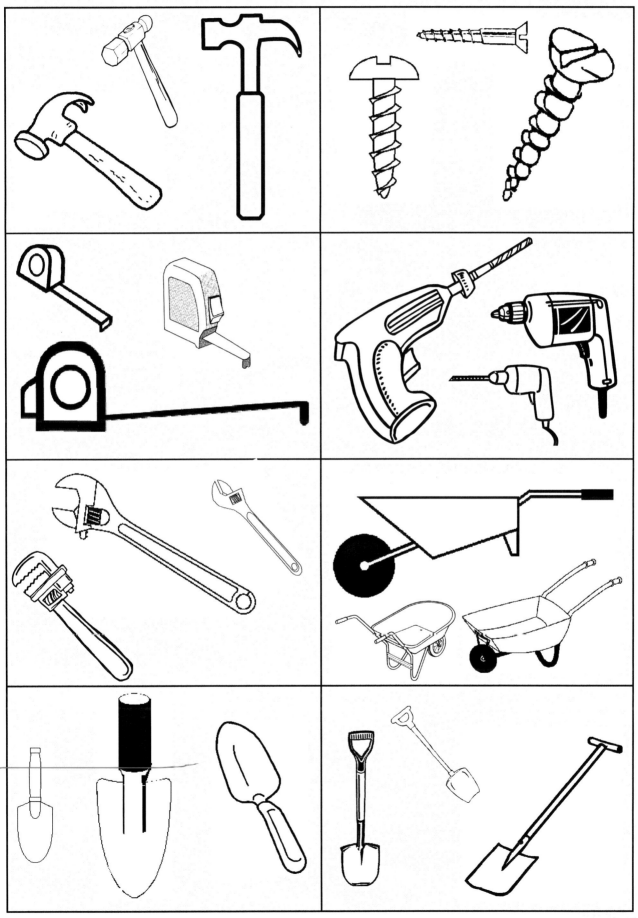

In each row colour the largest shape red, the smallest blue and the two the same size yellow.

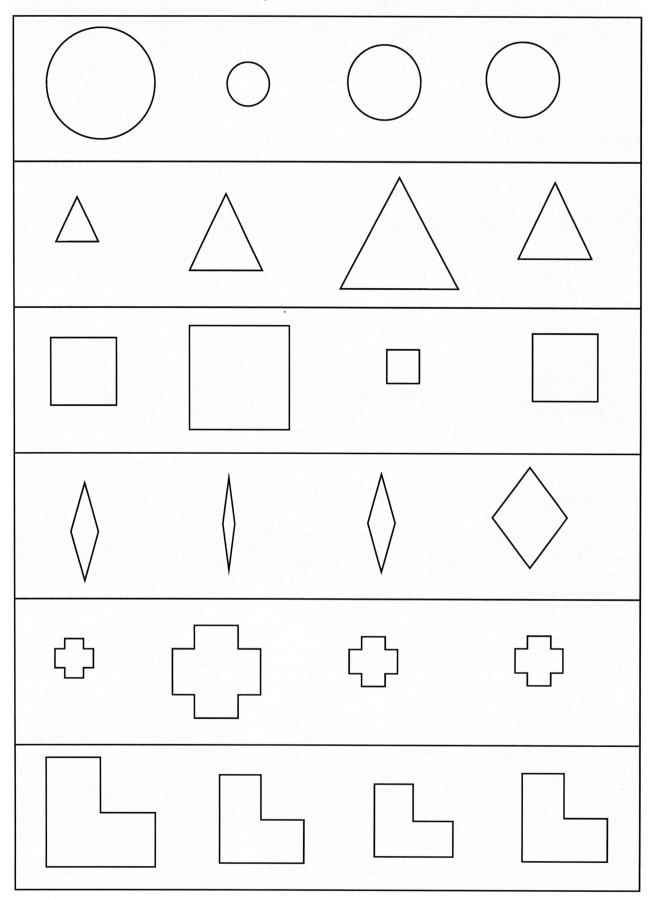

Circle the correct answer.

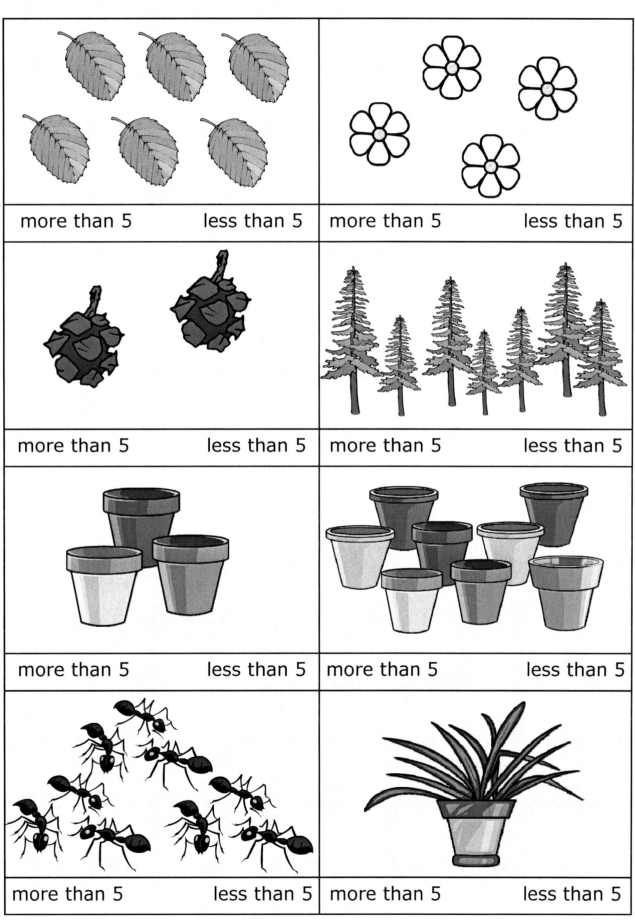

| more than 5 | less than 5 | more than 5 | less than 5 |

| more than 5 | less than 5 | more than 5 | less than 5 |

| more than 5 | less than 5 | more than 5 | less than 5 |

| more than 5 | less than 5 | more than 5 | less than 5 |

| N3.1 | Rounding numbers | 1b |

Complete the sequences.

| 1 | | 3 | | 5 | | 7 | | 9 | |

| | 2 | | 4 | | 6 | | 8 | | 10 |

| 1 | | | 4 | 5 | | | 8 | | 10 |

Circle the numbers in each row that are **larger** than the number in the box.

5	4	1	7	3	6	9	2
2	5	4	8	1	9	6	0
6	2	9	3	5	7	4	1

Write 4 numbers **smaller** than 8.

Write 4 numbers **smaller** than 10.

Place these numbers on the number line.

2   5   8   3   7

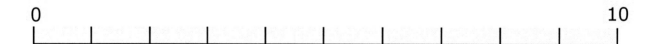

0   10

| N3.2 | Estimation | 1b |

Do not count but **estimate** which in each pair contains more.
Tick the box containing most.

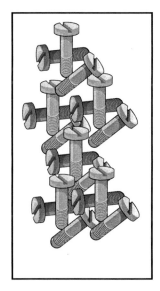 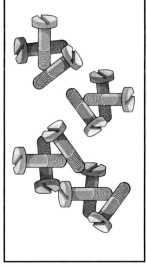

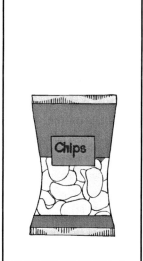

| **N3.3** | **Approximation** | **2c** |

When rounding to the nearest 10:

if the last number is 5 or above you round up

if the last number is less than 5 you round down.

$$71 \quad 72 \quad 73 \quad 74 \mid 75 \quad 76 \quad 77 \quad 78 \quad 79$$

These will round down to 70. | These will round up to 80.

Circle the numbers that would round down.

23   98   73   81   37   59   65   42   94   26   46   51

Circle the numbers that would round up.

53   78   26   55   94   24   18   83   16   82   49   37

Round these numbers down.

72 ⟶   51 ⟶   34 ⟶   21 ⟶

13 ⟶   82 ⟶   63 ⟶   44 ⟶

Round these numbers up.

79 ⟶   57 ⟶   36 ⟶   85 ⟶

68 ⟶   25 ⟶   17 ⟶   48 ⟶

Round up or down.

88 ⟶   51 ⟶   14 ⟶   75 ⟶

12 ⟶   47 ⟶   33 ⟶   76 ⟶

29 ⟶   64 ⟶   48 ⟶   22 ⟶

How many objects
in each set?

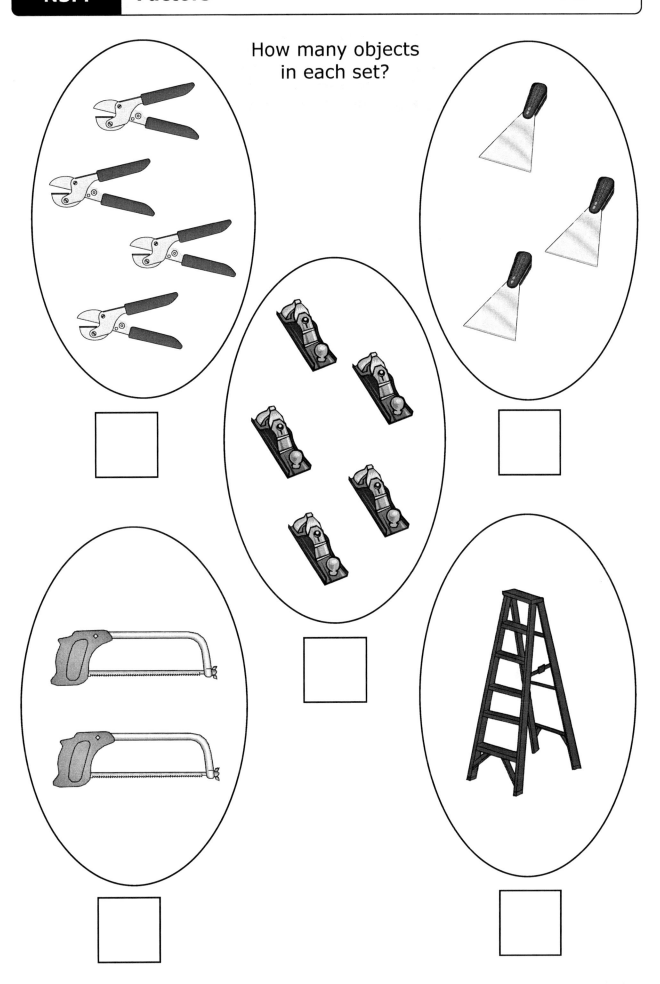

| **N3.4** | **Factors** | **1c** |

How many groups of two in each box?

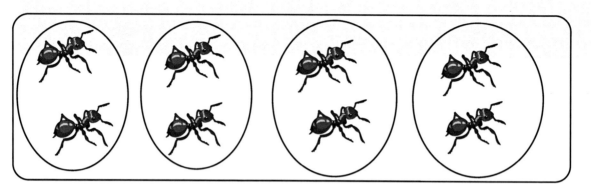

Circle the groups of objects.

| **N3.5** | **Multiples and factors** | **1c** |

Circle objects in groups.

groups of 2	
groups of 3	
groups of 4	
groups of 5	

| **N3.5** | **Multiples and factors** | **1b** |

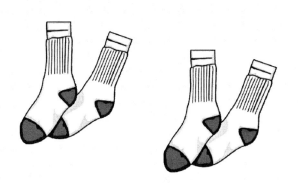

How many socks? ....
How many socks in a pair? ....
How many pairs of socks? ....

How many shoes? ....
How many shoes in a pair? ....
How many pairs of shoes? ....

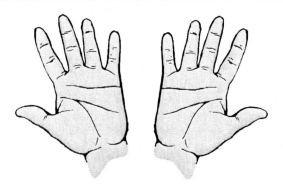

How many fingers? ....
How many fingers
on a hand? ......
How many hands? ....

How many bananas? ....
How many bananas in a bunch?
....
How many bunches of bananas?
......

How many pencils? ....
How many pencils
in a packet? ....
How many packets? ....

How many keys? ....
How many keys on a ring? ....
How many key rings? ....

Count the objects.

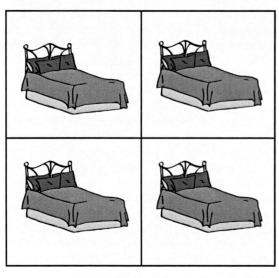

____ beds

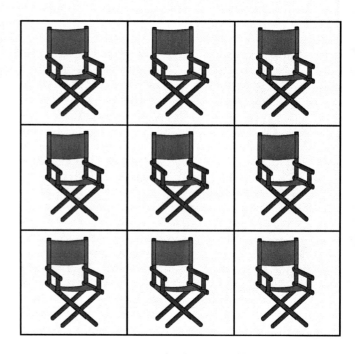

____ chairs

Draw a stool in each square

Draw the triangles.

MathsLinks

G2 2D shapes

| G2.1 | **Constructing triangles** | 1c |

Circle the odd shape out in each row.

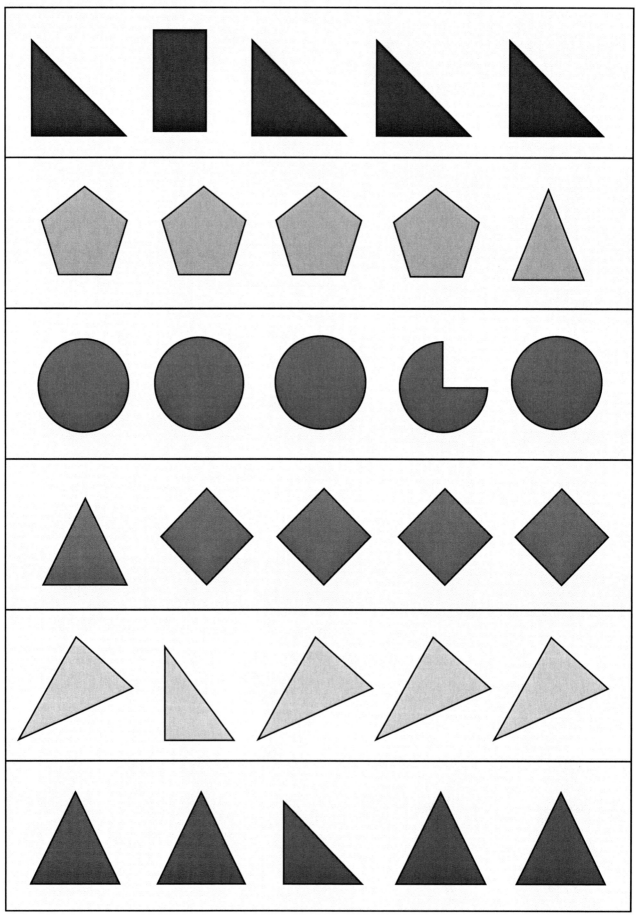

Complete these triangles and colour them.

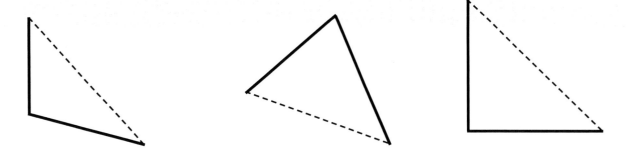

Colour the pattern.

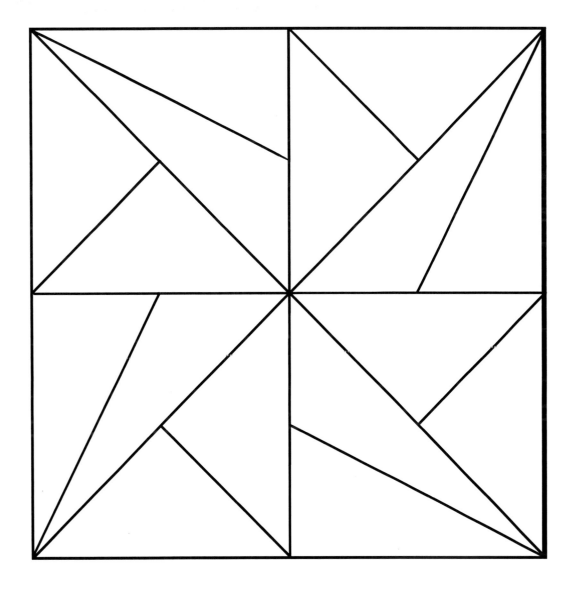

Circle the **largest** object in each row. Cross out the **smallest**.

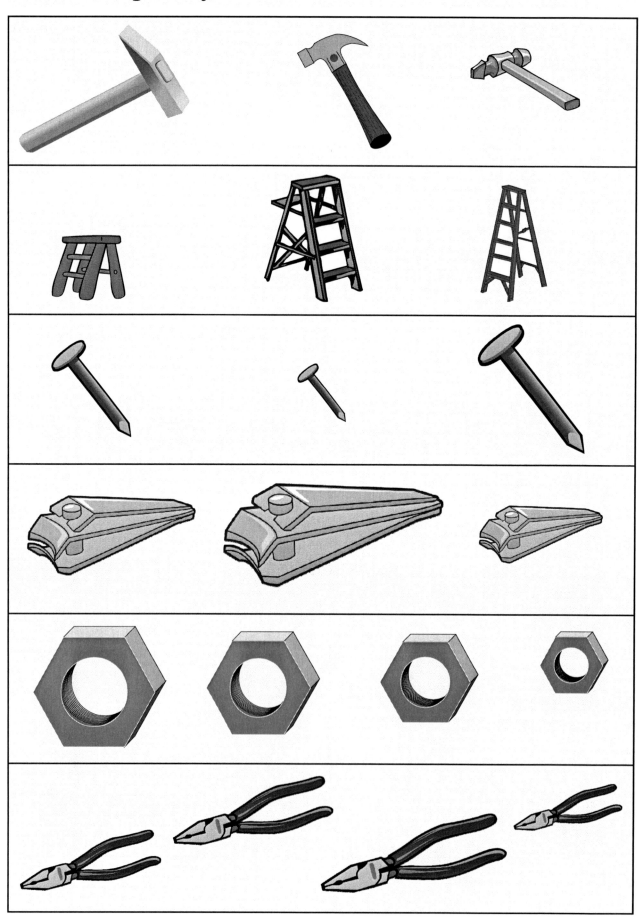

Draw round these shapes. This line is called the **perimeter**.

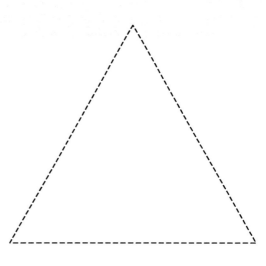

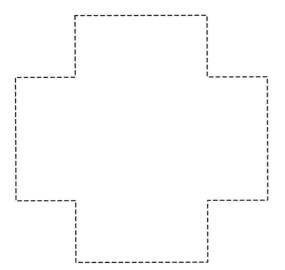
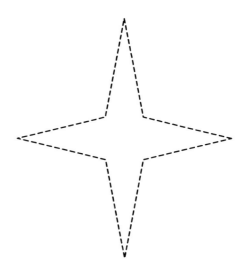

Draw the perimeter of the rectangles.

Draw a rectangle with a **smaller** perimeter than either and a

rectangle with a **larger** perimeter than either.

| **G2.3** | **Perimeter and area** | **1c** |

How many cubes long is each shape?

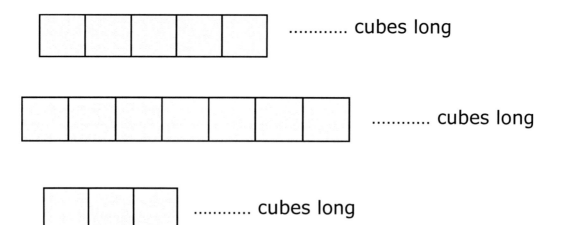

.... cubes long

.... cubes long

.... cubes long

How many cubes tall is each shape?

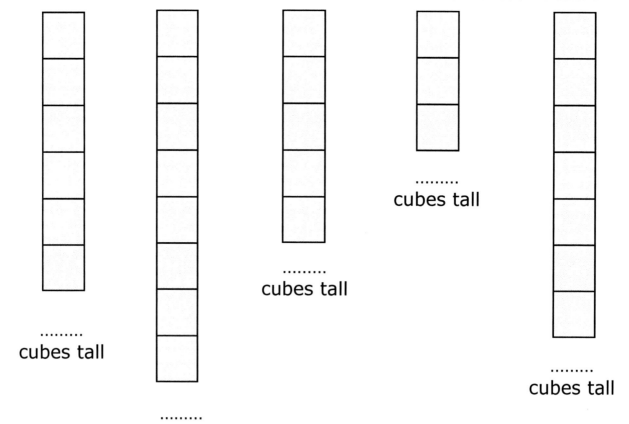

....
cubes tall

....
cubes tall

....
cubes tall

....
cubes tall

....
cubes tall

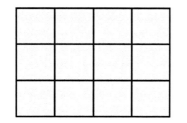

This shape is

.... cubes long

.... cubes tall

G2.3	Perimeter and area	1b

Read below each box then draw the shape in the box. They all have straight sides. You may need a ruler.

3 sides of different lengths

4 sides of different lengths

6 sides of different lengths

7 sides of different lengths

4 sides the same length

2 long sides and 2 short sides

Can you find the names of these shapes?

Colour the pattern of rectangles.

Colour the shapes.

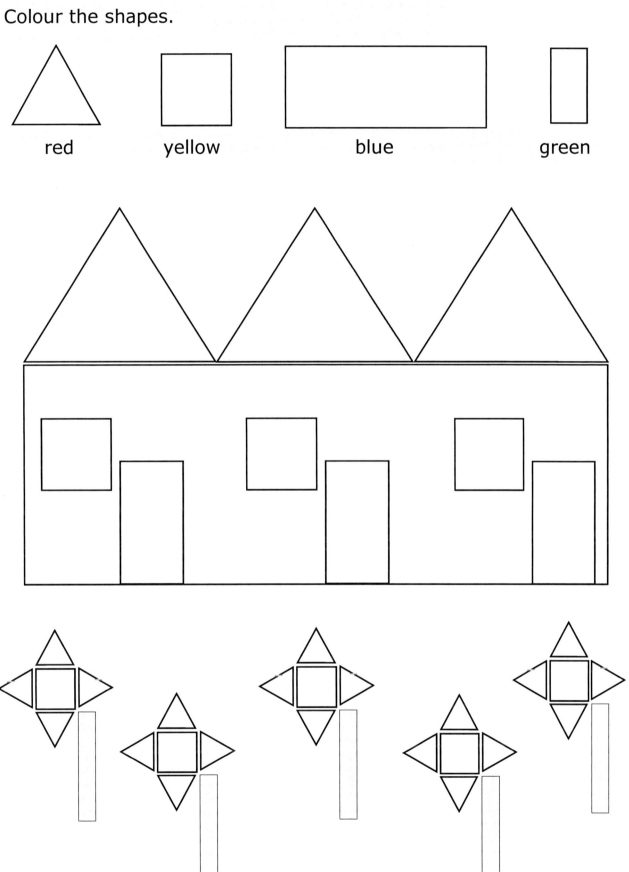

red   yellow   blue   green

This pattern is a quilting pattern made of rectangles.
Copy the pattern onto the grid.

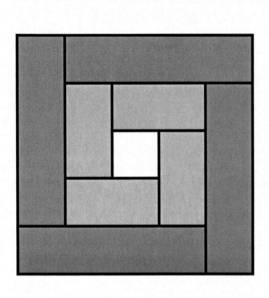

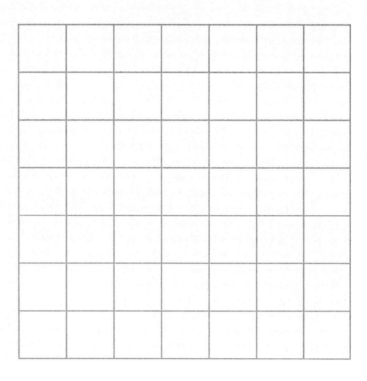

Continue this quilting pattern.
What happens if you turn it upside down?

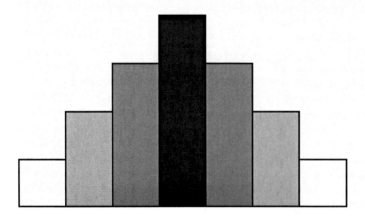

| D2.1 | **Drawing pictograms** | **P8** |

Draw the right number in each row.

4	
8	
9	
3	
7	
5	
1	

| D2.1 | **Drawing pictograms** | 1c |

How many are in each row?

	apples
	bananas
	peppers
	carrots
	mushrooms
	cobs of corn
	oranges
	lemons
	strawberries

There are the same number of carrots as ....

There are the same number of mushrooms as ....

One-to-One Mapping

Give each pencil a sharpener.

Give each envelope a stamp.

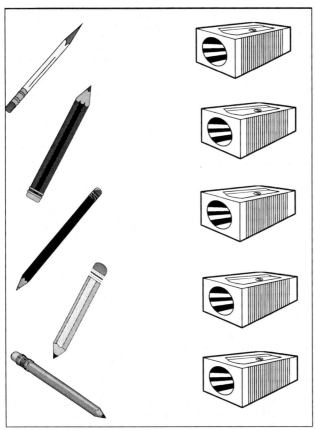

Give each mouse a mat.

Give each envelope a letter.

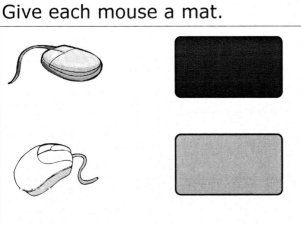

| D2.3 | **Selecting and drawing charts** | **P8** |

Draw the correct number in each row.

Draw 3 trees.	Draw 1 snake.
Draw a leaf.	Draw 4 cats.
Draw 2 flowers.	Draw 5 mice.

| **D2.3** | **Selecting and drawing charts** | **1c** |

Draw the correct number in each row.

3 ⬛	
1 ⬤	
2 ▬	
7 ▲	
4 ◇	
5 ◣	
8 ⬡	
6 ⌐	

Circle the groups.

How many in each group?

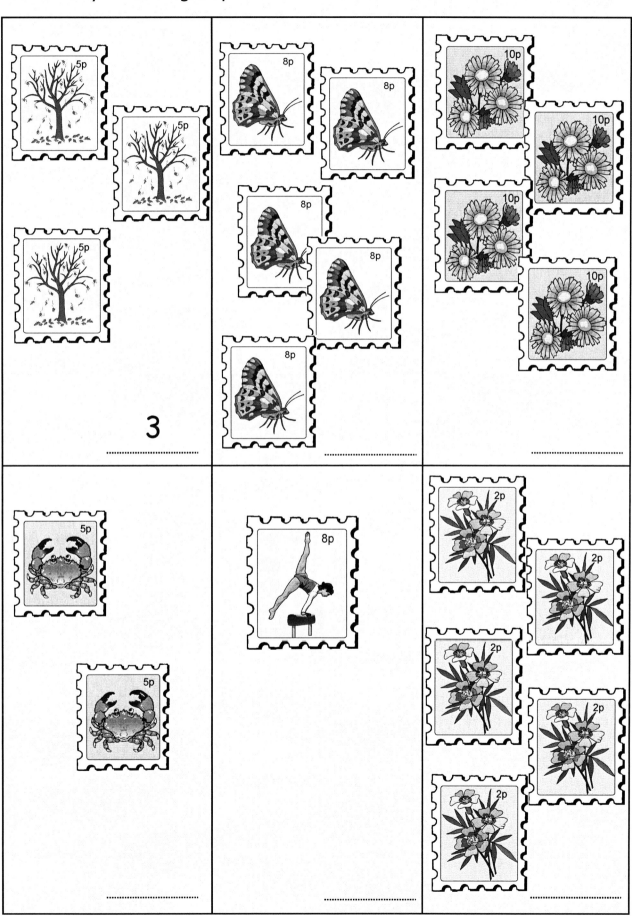

3

....

....

....

....

....

....

Divide each set into groups.

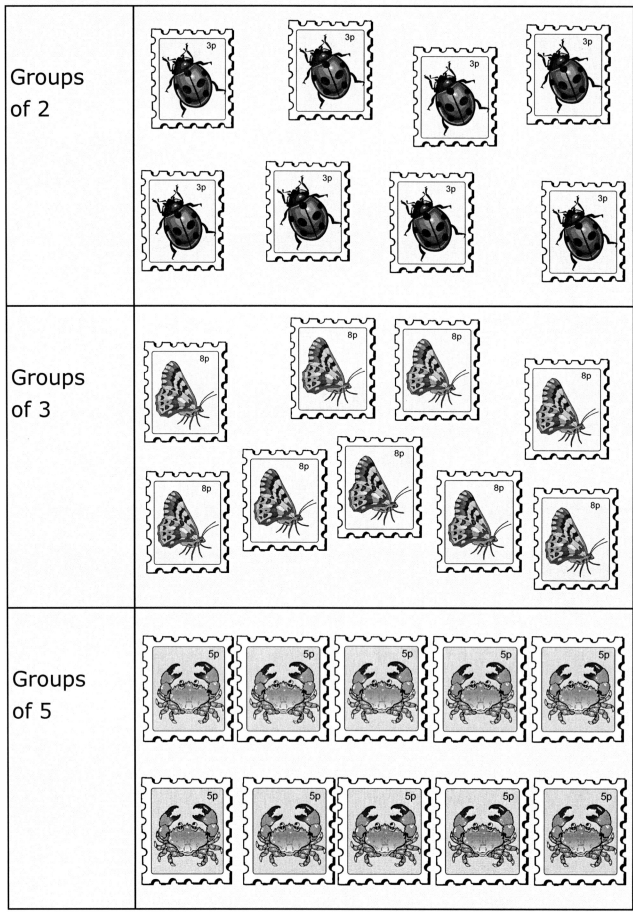

Groups of 2	
Groups of 3	
Groups of 5	

How many bricks in each row?

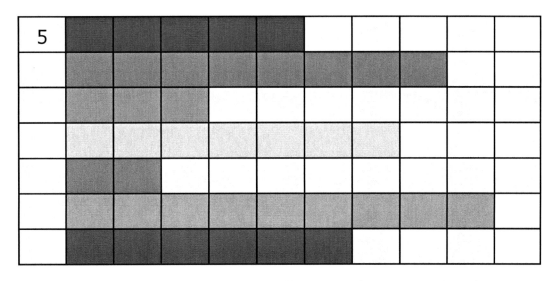

In each box colour the longest shape red and the shortest shape green.

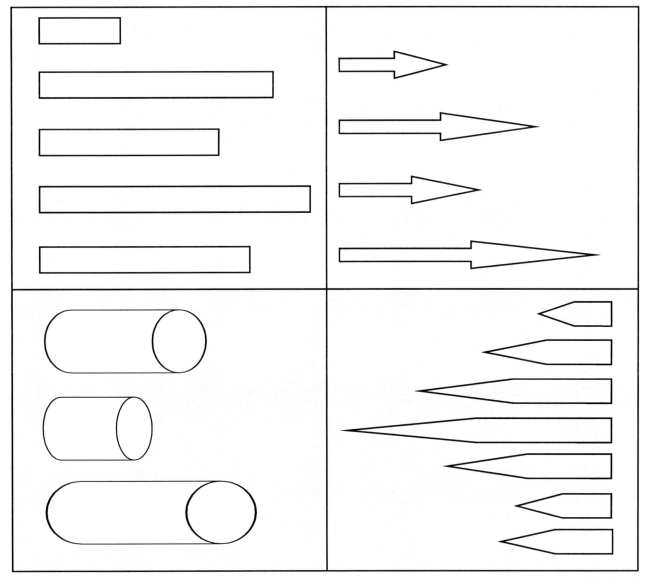

D2.5	Reading charts and diagrams	1c

These are objects found in the drawer.

labels

rubbers

brushes

scissors

sharpeners

pens

How many pens are there? ....

How many scissors are there? ....

There are .... brushes and .... rubbers.

What are there most of? ....

What are there fewest of? ....

Two objects have the same number. What are they? ....

Tick True or False.

	True	False
There are more brushes than rubbers.		
There are more scissors than labels.		
There are more sharpeners than pens.		
There are fewer pens than scissors.		
There are fewer rubbers than pens.		
There are fewer brushes than sharpeners.		

N4.1	Fractions	P8

How many parts are shaded?

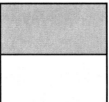   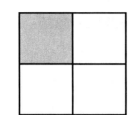   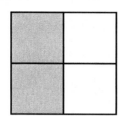

____ part out of ____   ____ part out of ____   ____ parts out of 4

Colour the correct number of parts.

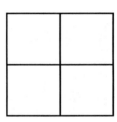   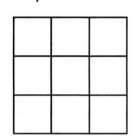   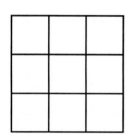

3 parts of 4   2 parts of 9   6 parts of 9

Write a statement to describe each line of bricks.

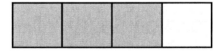   1 part of 4 is shaded.

   ____ part of ____ is shaded.

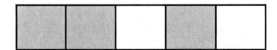

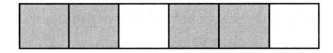

Count the number of equal parts each shape is divided into.

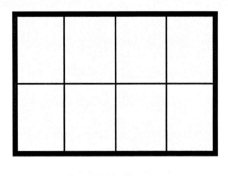

.... parts

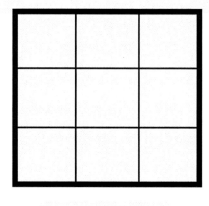

.... parts

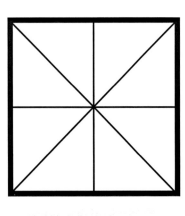

.... parts

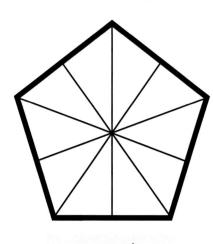

.... parts

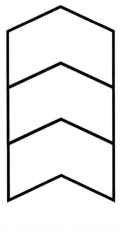

.... parts

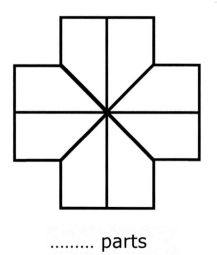

.... parts

Colour 1 part of each shape.

Link the objects that are the **same size**.

| N4.3 | **Fractions and decimals** | 2b |

This is 10p, or £0.10

Write the amount in each box in pence
and as a decimal fraction of £1.

40p   £0.40

N4.4   **Ordering decimals**   **P8**

What comes next?

N4.4	Ordering decimals	1c

Complete the number lines.

2 7

5 10

1 6

7 12

Circle the **larger** number in each box.

4	8	6	1	4	3
9	5	2	3	7	9
5	4	5	8	2	9
7	4	3	7	6	7

Which number comes **after**:

5 ....   7 ....   9 ....

2 ....   8 ....   1 ....

| **N4.5** | **Decimal numbers** | **2b** |

How much? Write the amounts in the line as part of a pound.

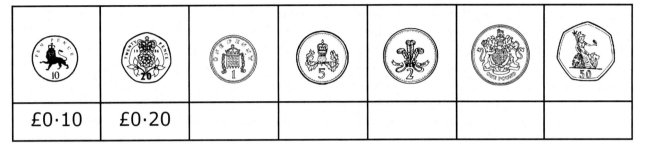

£0·10	£0·20					

Write these amounts in the boxes. The first one has been done for you.

£1· 25	£1	20p	5p	◯	◯
£1· 52	◯	◯	◯	◯	◯
£1· 34	◯	◯	◯	◯	◯
£2 · 51	◯	◯	◯	◯	◯
£2 · 70	◯	◯	◯	◯	◯
£1· 67	◯	◯	◯	◯	◯

| **N4.6** | **Understanding decimals** | **2c** |

Find the sums.

20 + 10 = <u>30</u> + 10 = <u>40</u> + 10 = <u>50</u> + 10 = 60

80 - 10 = ___ - 10 = ___ - 10 = ___ - 10 = ___

90 - 10 = ___ - 10 = ___ - 10 = ___ - 10 = ___

40 + 10 = ___ + 10 = ___ + 10 = ___ + 10 = ___

50 - 10 = ___ - 10 = ___ - 10 ___ - 10 = ___

35 + 10 = ___ + 10 = ___ + 10 = ___ + 10 = ___

Put the correct number of beads on the abacus.

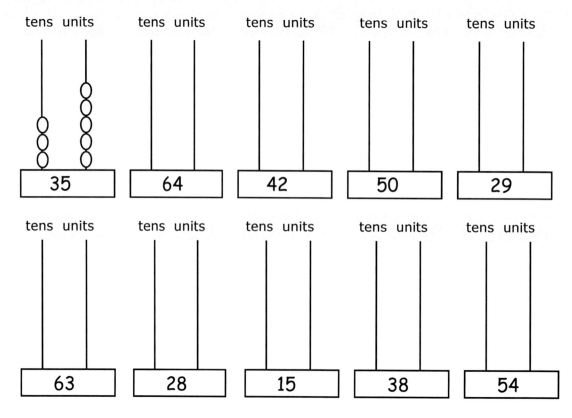

How much money in each box?

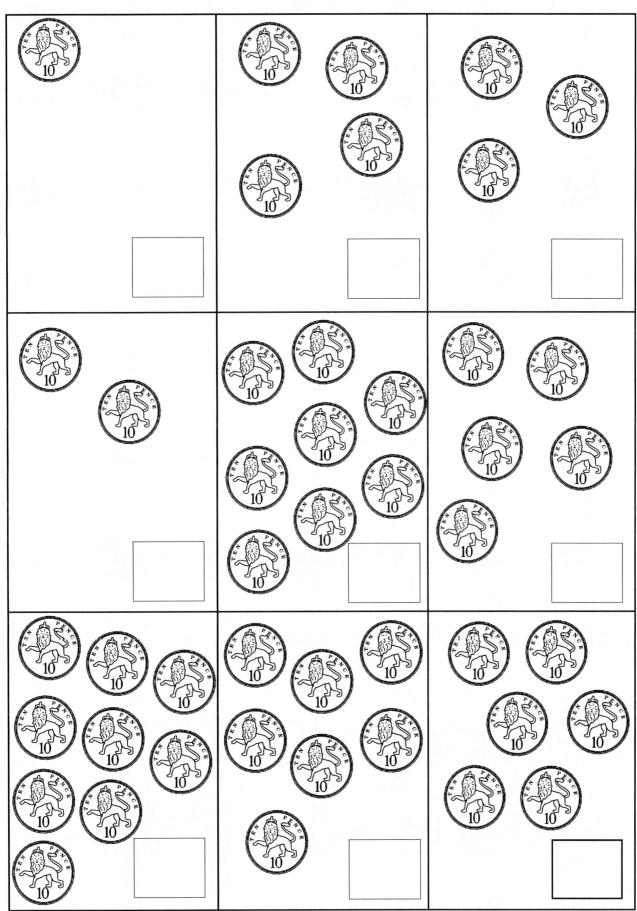

How much? ....
Add £0.10
How much now? ....

How much? ....
Add £0.10
How much now? ....

How much? ....
Add £0.10
How much now? ....

How much? ....
Add £0.10
How much now? ....

How much? ....
Add £0.20
How much now? ....

How much? ....
Add £0.20
How much now? ....

How much? ....
Add £0.20
How much now? ....

How much? ....
Add £0.30
How much now? ....

| G3.1 | 3-D shapes | P7 |

Draw the shapes.

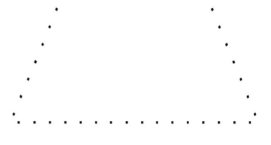

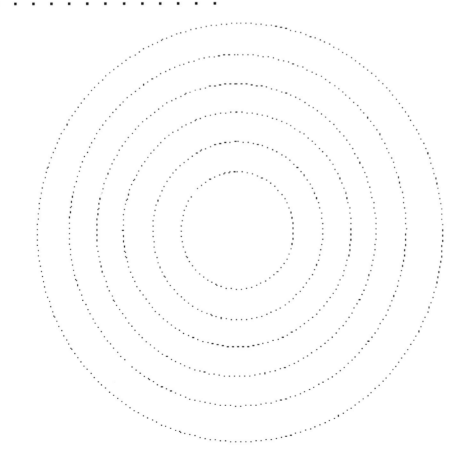

Colour the shapes.

G3.1 | **3-D shapes** | **1c**

Circle the odd one out in each row.

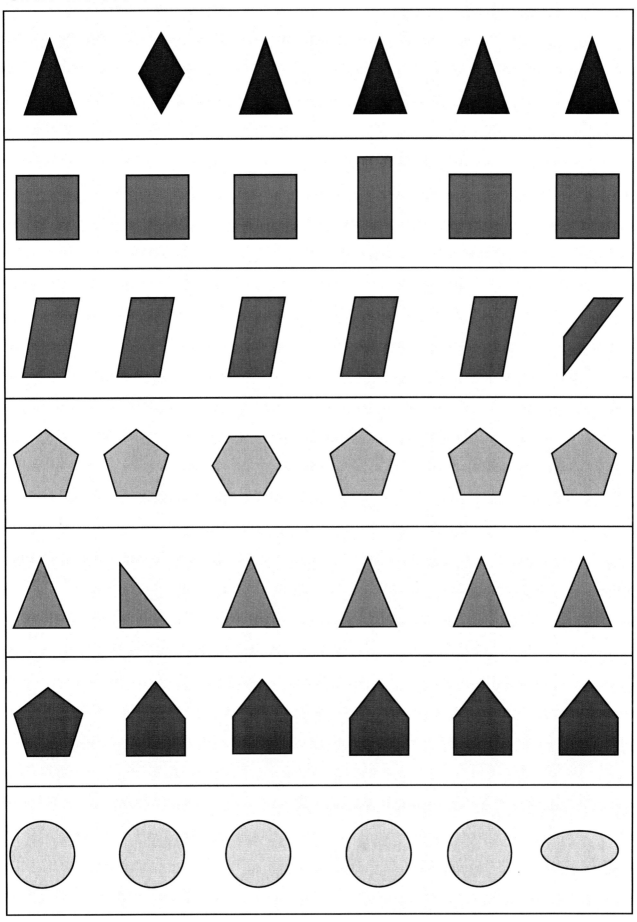

113

| G3.2 | Describing 3-D shapes | 1c |

Circle the matching shape.

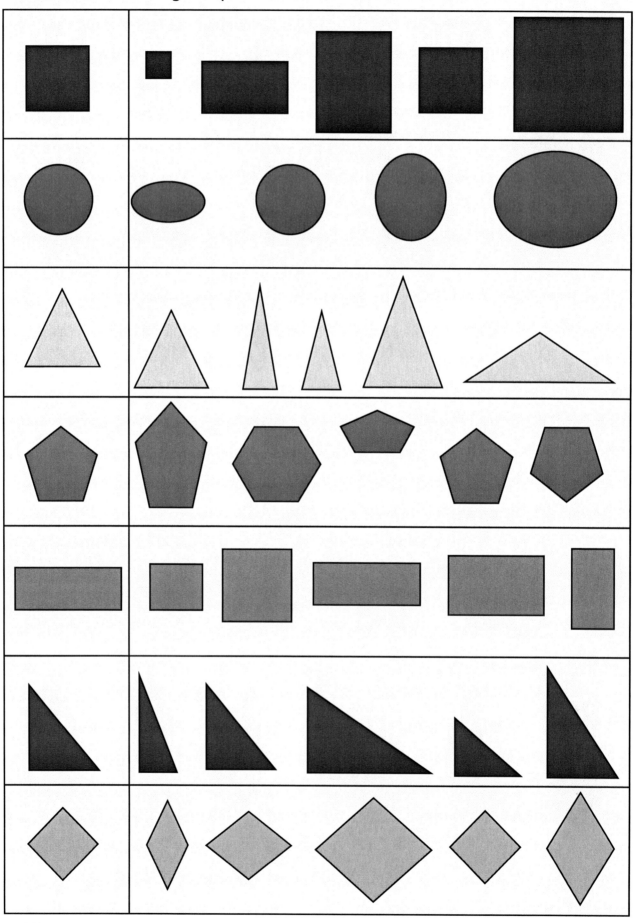

| **G3.3** | **Solid shapes and surface area** | **1b** |

Match the toys and the shapes.

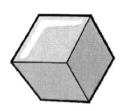

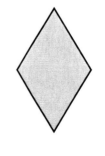

Fill in the table below.

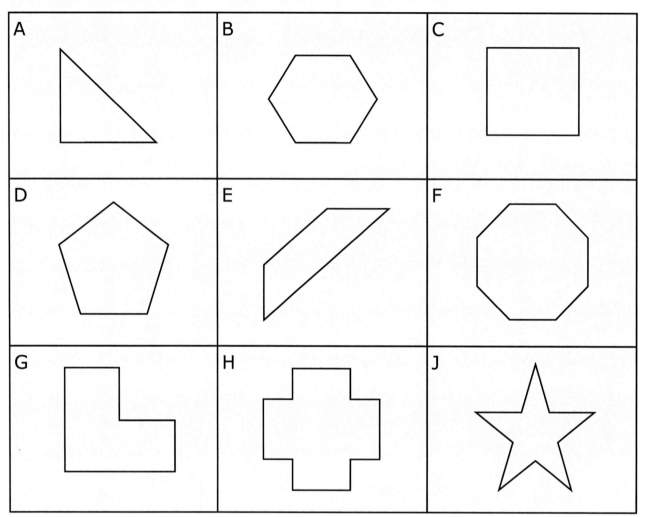

Shape	Sides	Corners
A	3	3
B		
C		
D		
E		
F		
G		
H		
J		

| **G3.4** | **Surface area of a cuboid** | **1a** |

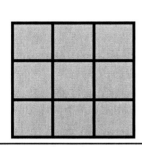

| length = 3 |
| width = 3 |
| area =....9.... squares |

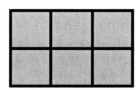

| length = |
| width = |
| area =......... squares |

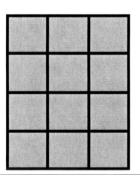

| length = |
| width = |
| area =......... squares |

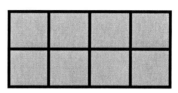

| length = |
| width = |
| area =......... squares |

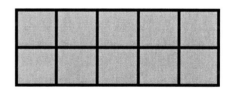

| length = |
| width = |
| area =......... squares |

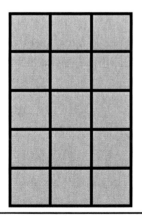

| length = |
| width = |
| area =......... squares |

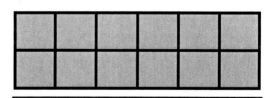

| length = |
| width = |
| area =......... squares |

G3.5	Volume	1c

How many boxes in each picture?

1 2 3	3 4 5	1 2 3
5 6 7	1 2 3	3 4 5
2 3 4	5 6 7	Draw 5 boxes.

| **G3.5** | **Volume** | **1a** |

Circle the correct description for each object.

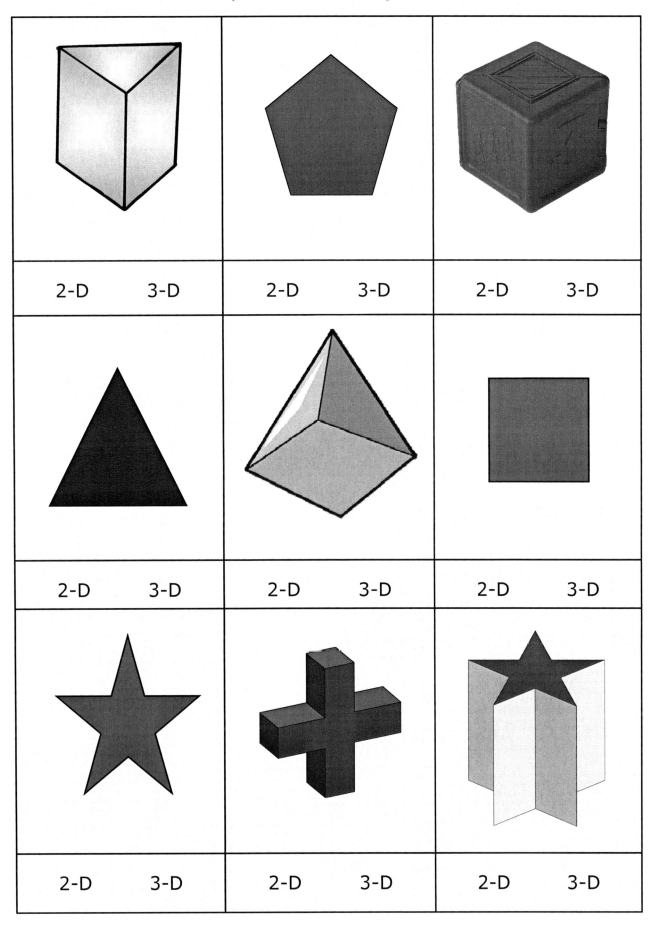

| 2-D | 3-D | 2-D | 3-D | 2-D | 3-D |

| 2-D | 3-D | 2-D | 3-D | 2-D | 3-D |

| 2-D | 3-D | 2-D | 3-D | 2-D | 3-D |

| G3.6 | Volume of a cuboid | 1a |

How many squares in the rectangles?

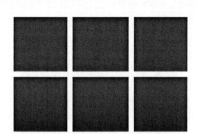

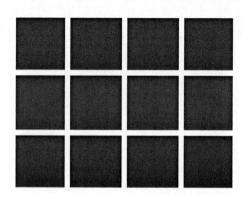

How many cubes in each cuboid?

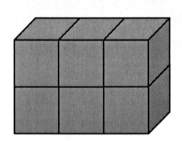

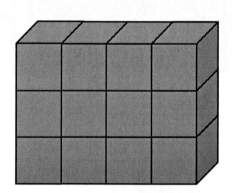

How many cubes in the 2-by-2-by-1 cuboid
on the isometric paper below? ....
Draw a 3-by-3-by-1 cuboid on the isometric paper.

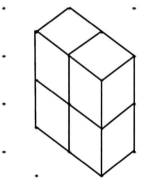

One-to-one mapping

Link each racket to a shuttlecock.

Give each person a whistle.

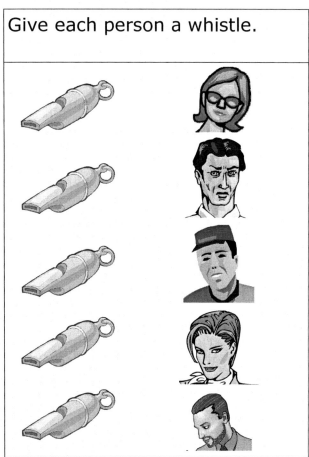

Give each child a hat.

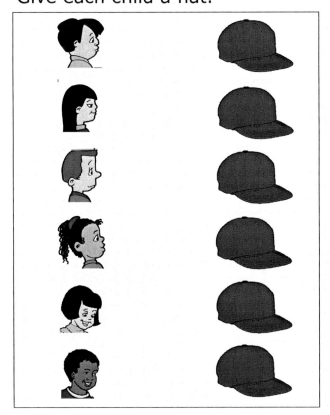

Link each pair of socks to a boot.

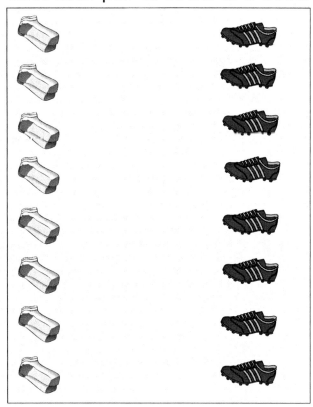

Tick the things you can eat and cross out those you cannot eat.

D3.2 | **Possibility** | **1a**

Read the sentences and decide whether they are true or false.

A book has pages.	✓
All flowers are green.	✗
Water is wet.	
Trees have leaves.	
Wheels are square.	
All girls have long hair.	
Cars have wheels.	
Gloves are for hands.	
Hats go on heads.	
Shoes go on ears.	
Everyone likes carrots.	
Carrots are orange.	
Baby cats are kittens.	
Baby cows are piglets.	
Hens lay eggs.	

Circle the matching object.

What will they grow into? Join them with lines.

| **D3.3** | **The probability scale** | **1b** |

Tick the Yes or No box

	Yes	No
We get milk from cows.		
We get eggs from rabbits.		
A young dog is called a puppy.		
Footballers use a bat.		
It is dark at night.		
Apples can be red.		
Bananas are blue.		
Leaves are green.		
Cats have two eyes and two ears.		
Dogs say moo.		
This month is May.		
Everyone has a bicycle.		
Snap is a card game.		
A car needs a driver.		
You wear a hat on your feet.		

Link the pictures that are the same.

D3.4   **Understanding probability**   1c

Find the odd one out.

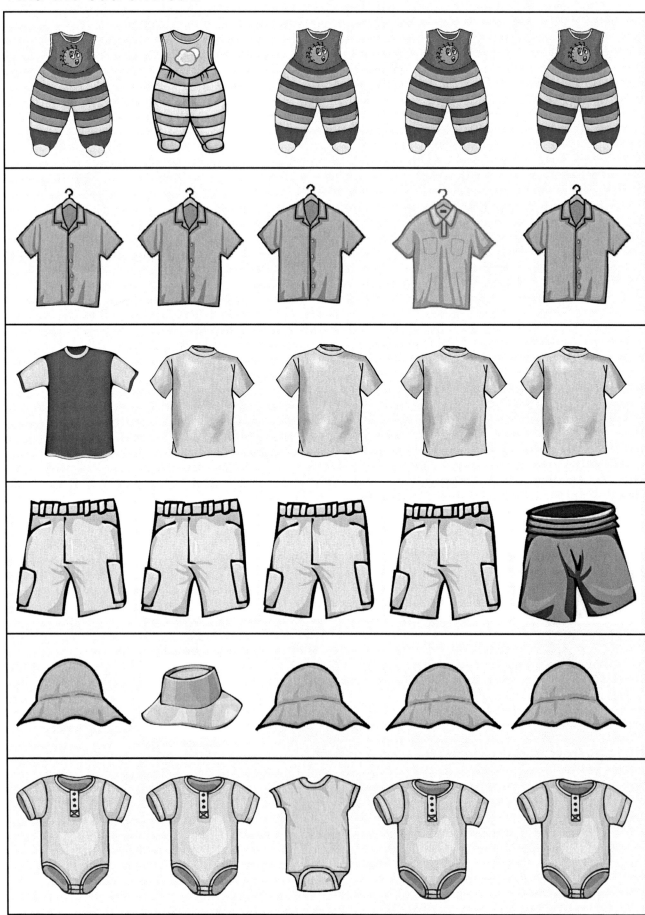

D3.5   **Equivalent probabilities**   **P8**

Equivalent means the **same**. Select the equivalent amount.

1 2 3	3 4 5	6 7 8
3 4 5	1 2 3	2 3 4
3 4 5 6	1 2 3 4	1 2 3 4

| **D3.5** | **Equivalent probabilities** | **1c** |

Colour the correct number of sections for each fan.

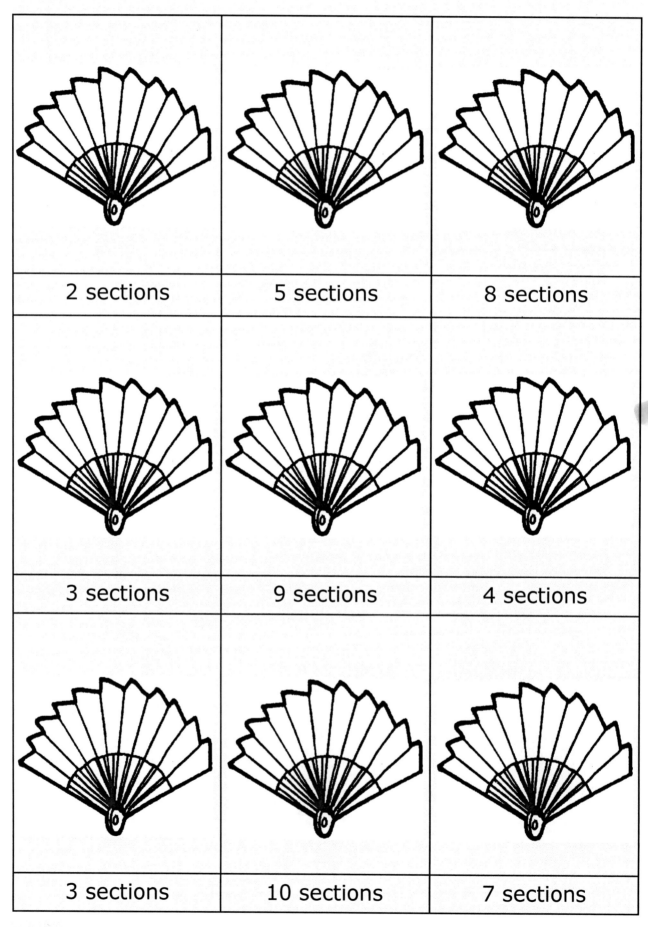

2 sections	5 sections	8 sections
3 sections	9 sections	4 sections
3 sections	10 sections	7 sections

Computers can number the pages of a book.

If this page is 12, the next page will be 13.

If this page is 9, the next page will be ____ .

If this page is 19, the next page will be ____ .

If this page is 18, the next page will be ____ .

The next page is 4, so this page is ____ .

The next page is 11, so this page is ____ .

Look at a calculator and colour in how it displays these numbers.

2	7	4
3	9	5

Each machine will add 1 or subtract 1.

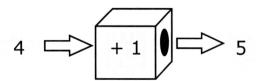

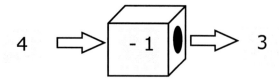

Can you work out what each machine is doing?

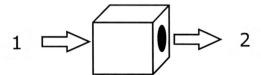

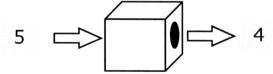

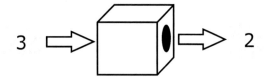

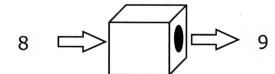

Find the output for each machine when the input is 5.

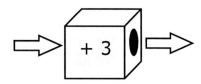

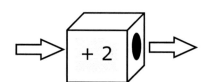

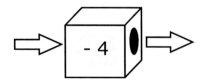

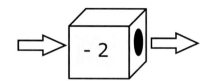

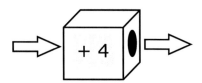

A4.2	Doing and undoing	P7

Draw a **smaller** circle.   Draw a **larger** circle.

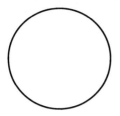

Draw a **smaller** triangle.   Draw a **larger** triangle.

Colour all the arrows pointing **up**.

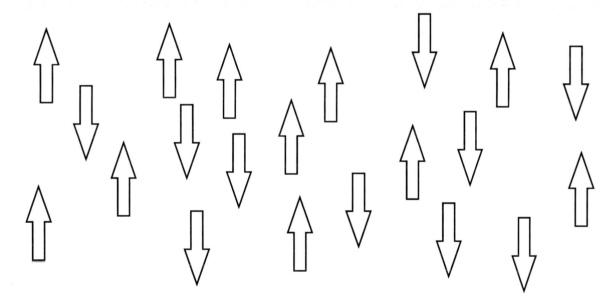

1   2   3   4   5

Use the numbers above to finish the line below.

5   4   ___   ___   ___

Link the pictures that show **opposites**.

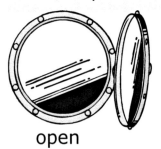

open

fast

empty

long

cold

new

slow

short

hot

closed

old

full

Circle the 2 that are the same size in each row.

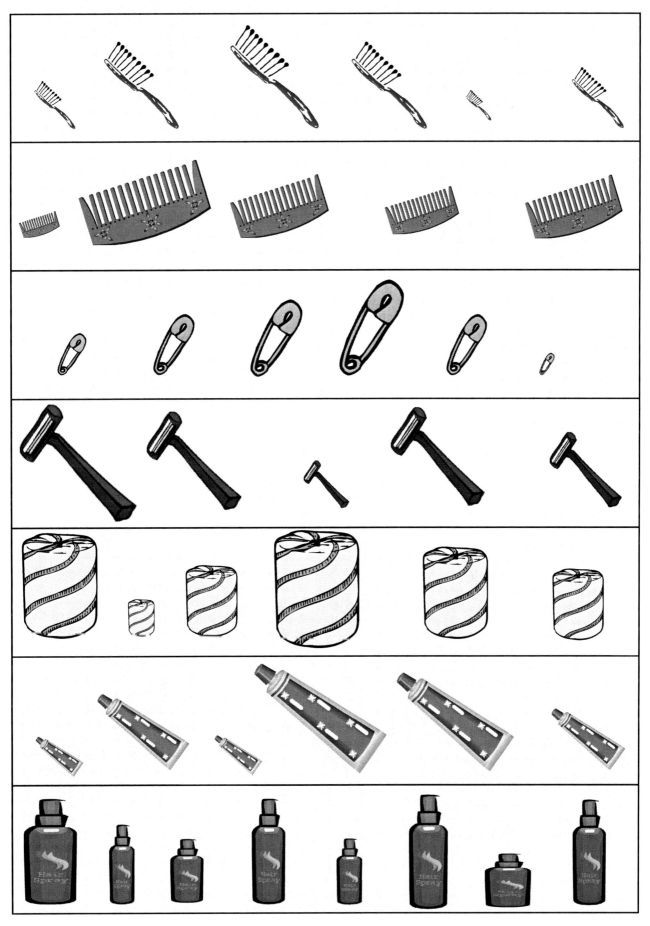

Link those with the same value.

one
two
three
four
five
six

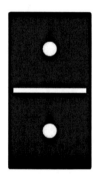

Opposites
Link the pictures that show opposites.

| A4.4 | Inverse operations | 1c |

Addition and Subtraction

How many pliers? ....

Draw one more.

How many pliers now? ....

How many pliers? ....

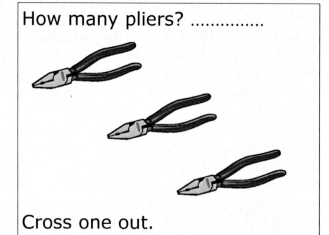

Cross one out.

How many pliers now? ....

How many bolts? ....

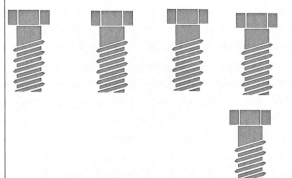

Draw one more.

How many bolts now? ....

How many bolts? ....

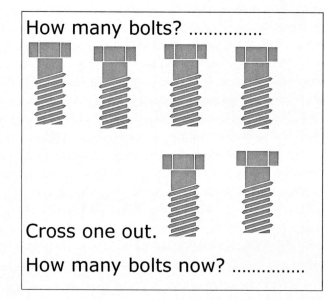

Cross one out.

How many bolts now? ....

How many locks? ....

Draw two more.

How many locks now? ....

How many locks? ....

Cross two out.

How many locks now? ....

Select the correct action.

pull push	up down
crying laughing	sitting standing
asleep awake	fast slow
below above	young old

How many screws? ....

Draw one more.

How many screws now? ....

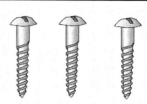

How many screws? ....

Cross one out.

How many screws now? ....

How many nails? ....

Draw one more.

How many nails now? ....

How many nails? ....

Cross one out.

How many nails now? ....

How many bolts? ....

Draw one more.

How many bolts now? ....

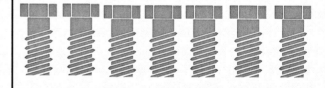

How many bolts? ....

Cross one out.

How many bolts now? ....

How many nuts? ....

Draw one more.

How many nuts now? ....

How many nuts? ....

Cross one out.

How many nuts now? ....

Colour the pattern remembering the symmetry properties.

This square has 4 lines of symmetry.

The pattern is the diagram of a cube rotated 4 times.

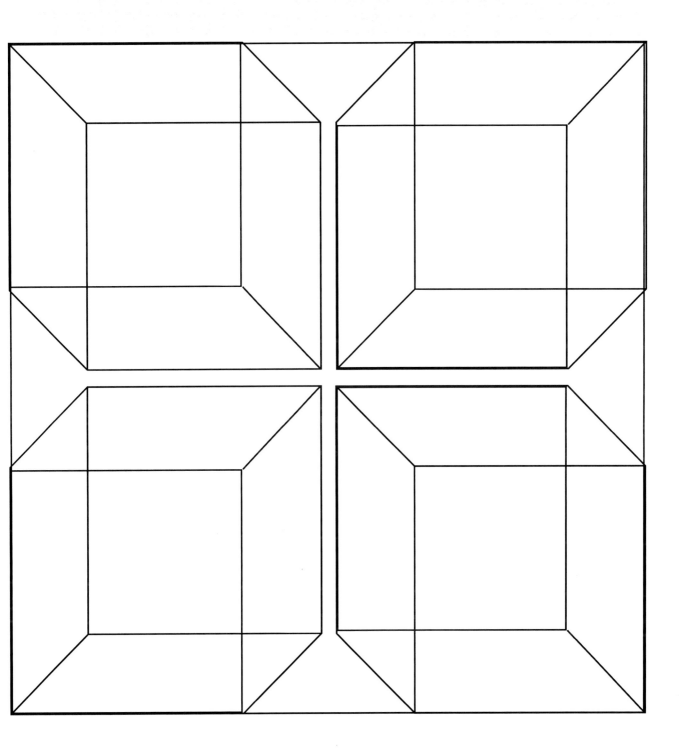

Draw the lines of symmetry on these pictures.

The picture shows each side of a fish.

Colour the fish so it is symmetrical.

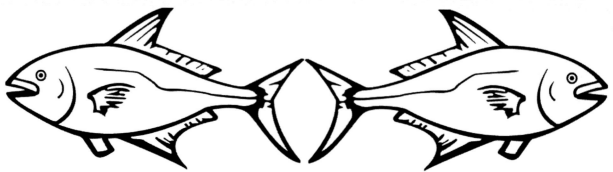

G4.2 | **Tessellations** | **1c**

Complete the patterns.

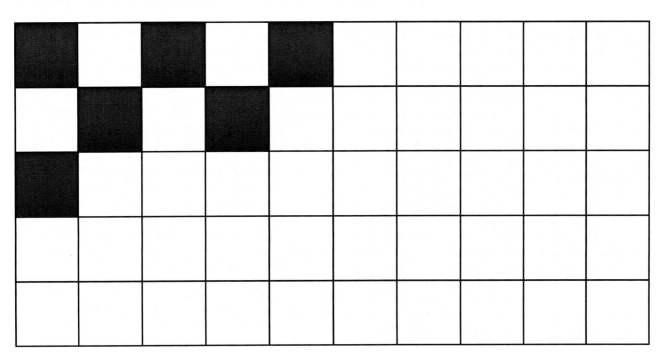

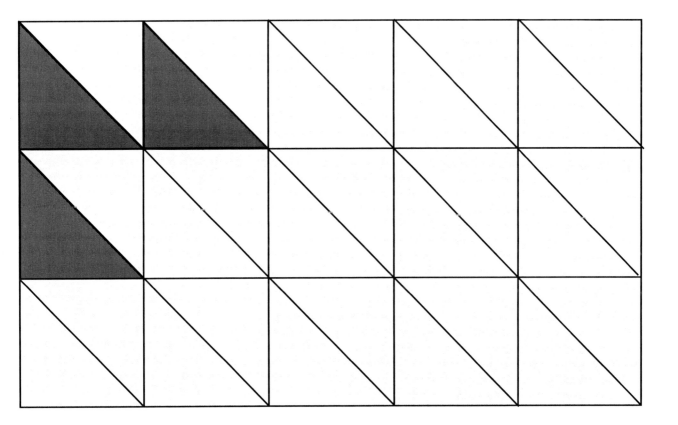

A **tessellation** is a pattern that repeats.
This shape can be tessellated in several ways.

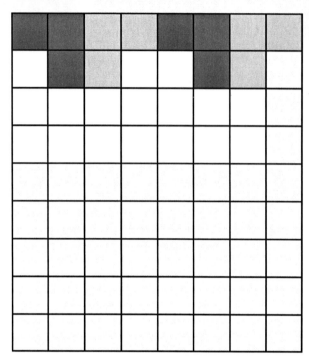

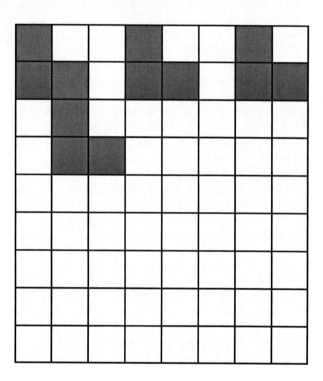

Create 2 more tessellations using the same shape.

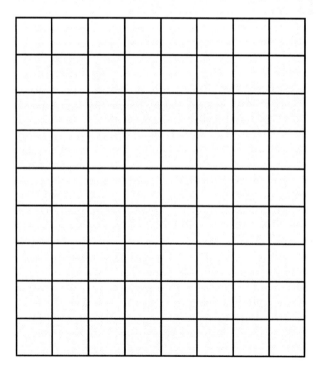

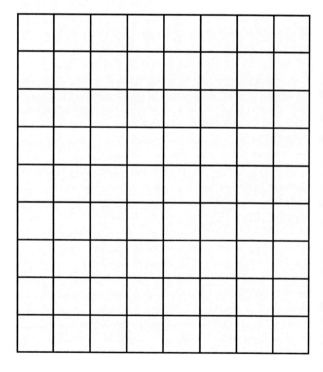

Draw the next shape in each row.

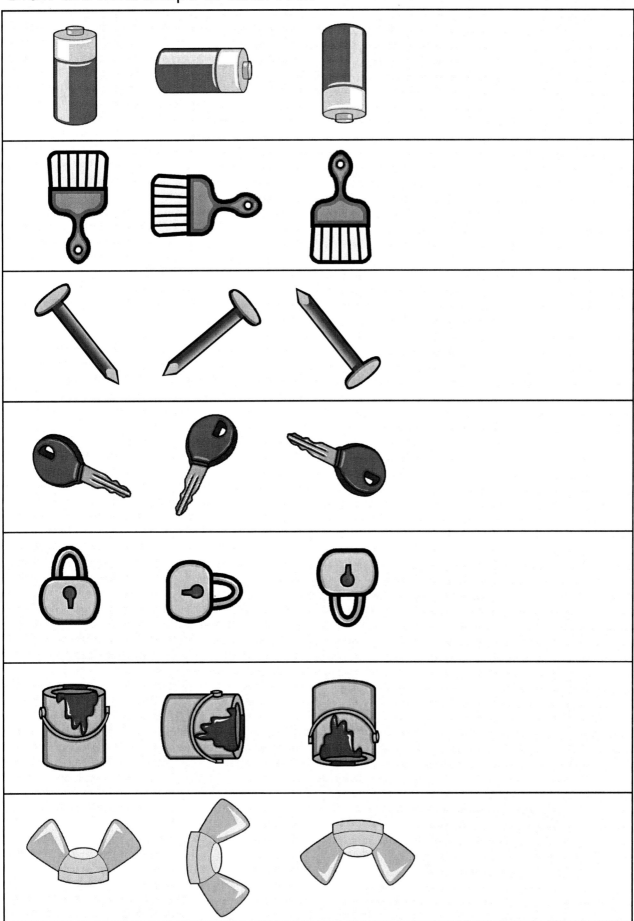

Draw the next shape in each row.

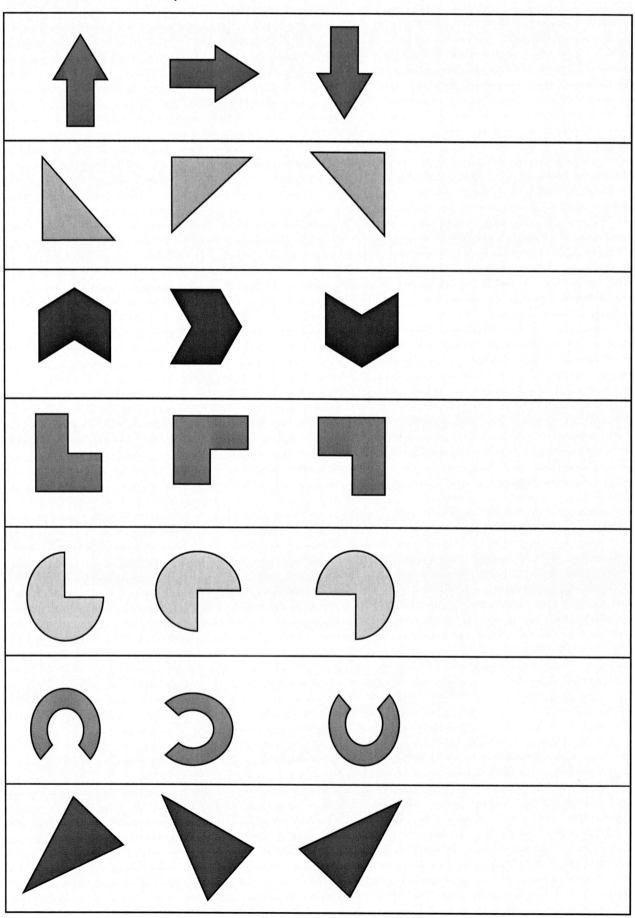

Mark the lines of symmetry in each picture.

Translate this shape to 5 different places on the grid.

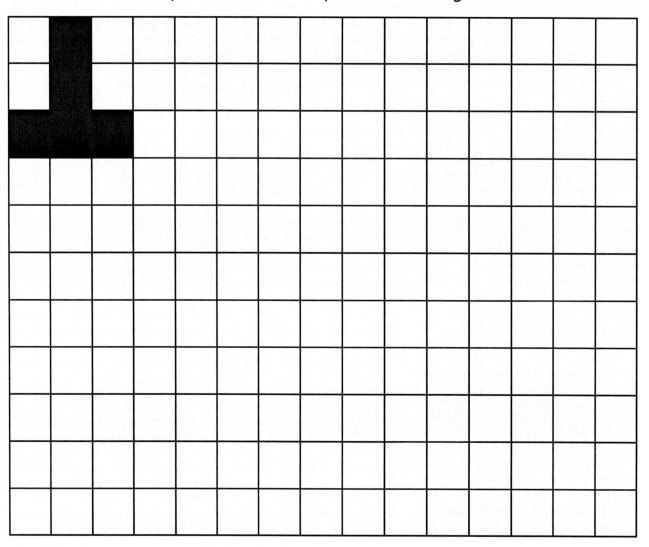

Draw the reflections of these shapes to make a new larger shape.

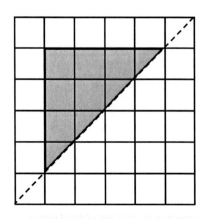

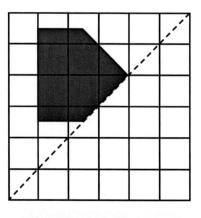

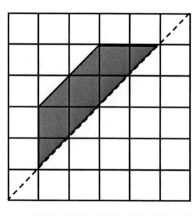

The new shape

has .... sides

The new shape

has .... sides

The new shape

has .... sides

Colour the **largest** shape in each row red.
Colour the **smallest** shape in each row blue.

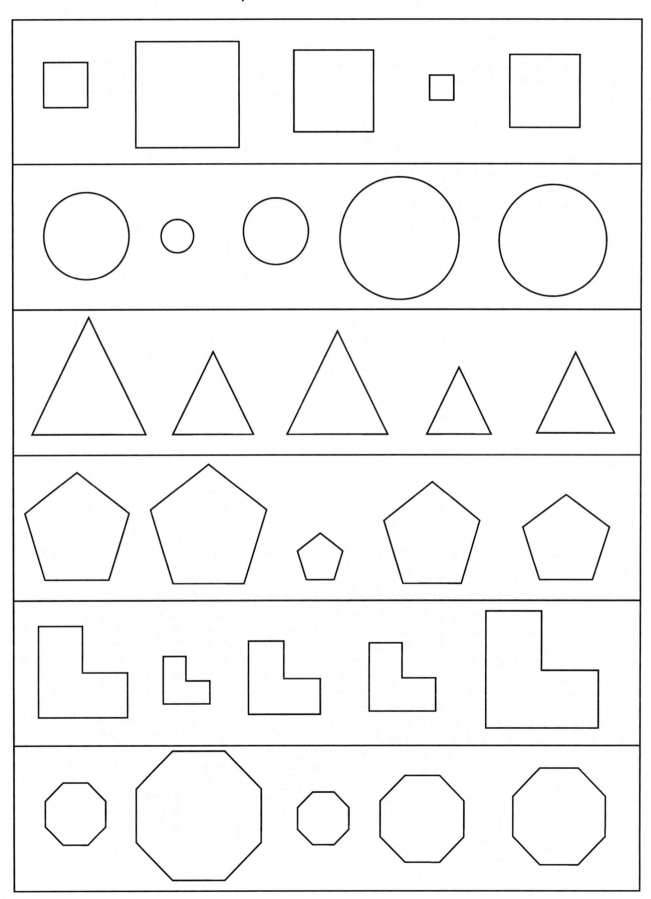

Draw a **larger** shape in each box.

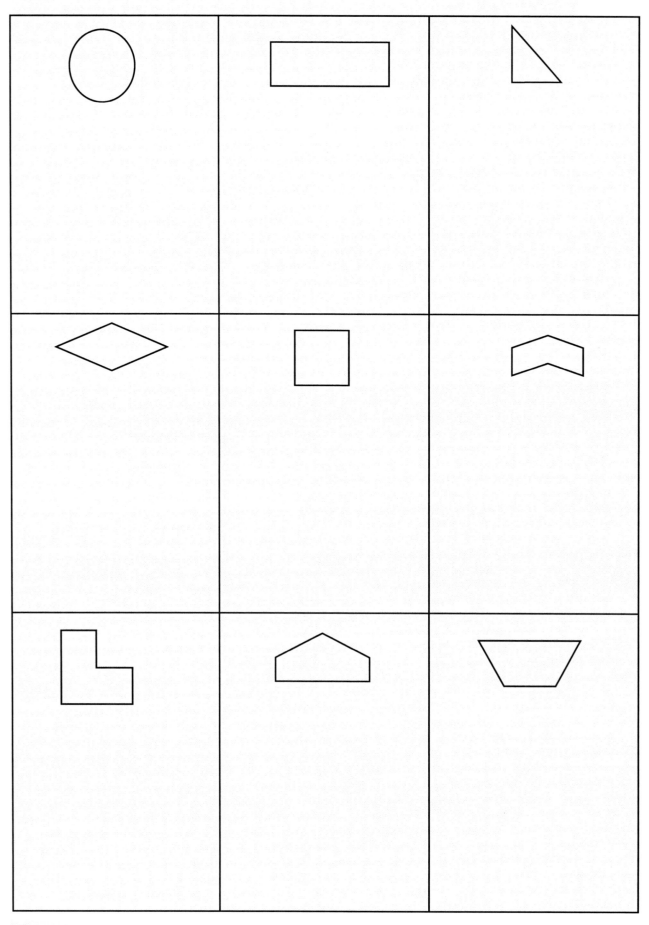

N5.1	Fractions of amounts	1b

Share the sweets between the 3 plates.

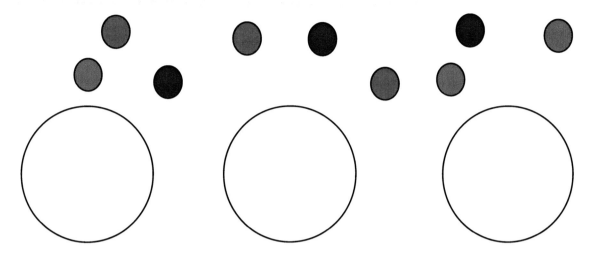

How many sweets altogether? …………

How many sweets on each plate? …………

Share the balls between the 4 bags.

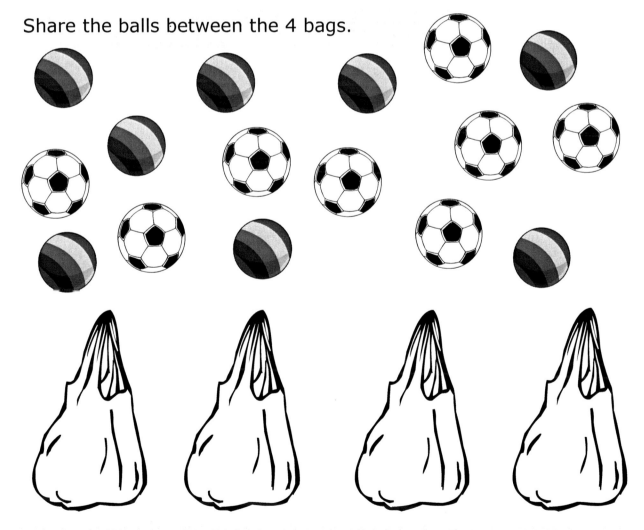

How many balls altogether? …… How many balls in each bag? ……

N5.1	**Fractions of amounts**	**1a**

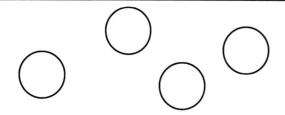

There are .....4..... circles.

Colour half of the circles green.

Half of .....4..... is .....2.....

There are .... pencils.

Colour half of the pencils red.

Half of .... is ....

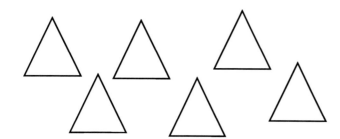

There are .... triangles.

Colour half of the triangles blue.

Half of .... is ....

There are .... leaves.

Colour half of the leaves green.

Half of .... is ....

There are .... squares.

Colour half of the squares yellow.

Half of .... is ....

There are .... cakes.

Colour half of the cakes green.

Half of .... is ....

| **N5.2** | **Ratio** | **P8** |

There are .... pins
and .... clips.

There are .... envelopes
and .... pencils.

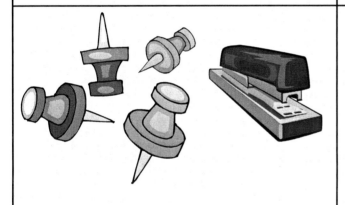

There are .... pins
and .... stapler.

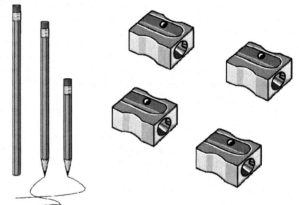

There are .... pencils
and .... sharpeners.

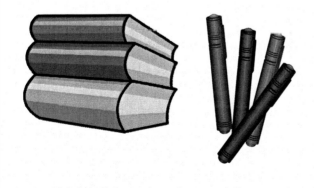

There are .... books
and .... pens.

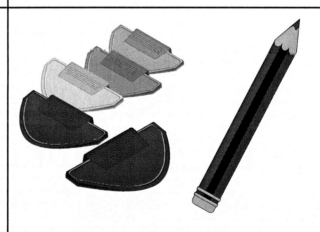

There are .... clips
and .... pencil.

| **N5.2** | **Ratio** | **1c** |

Colour 2 red Colour 1 blue Ratio 2 : 1	
Colour 2 green Colour 3 blue Ratio 2 : 3	
Colour 1 yellow Colour 4 red Ratio 1 : 4	
Colour 2 purple Colour 4 orange Ratio 2 : 4	
Colour 1 blue Colour 3 green Ratio 1 : 3	
Colour 5 orange Colour 3 brown Ratio 5 : 3	
Colour 2 yellow Colour 5 red Ratio 2 : 5	
Colour 3 blue Colour 4 orange Ratio 3 : 4	

N5.3 **Proportion** **P7**

Circle groups of 2 triangles.

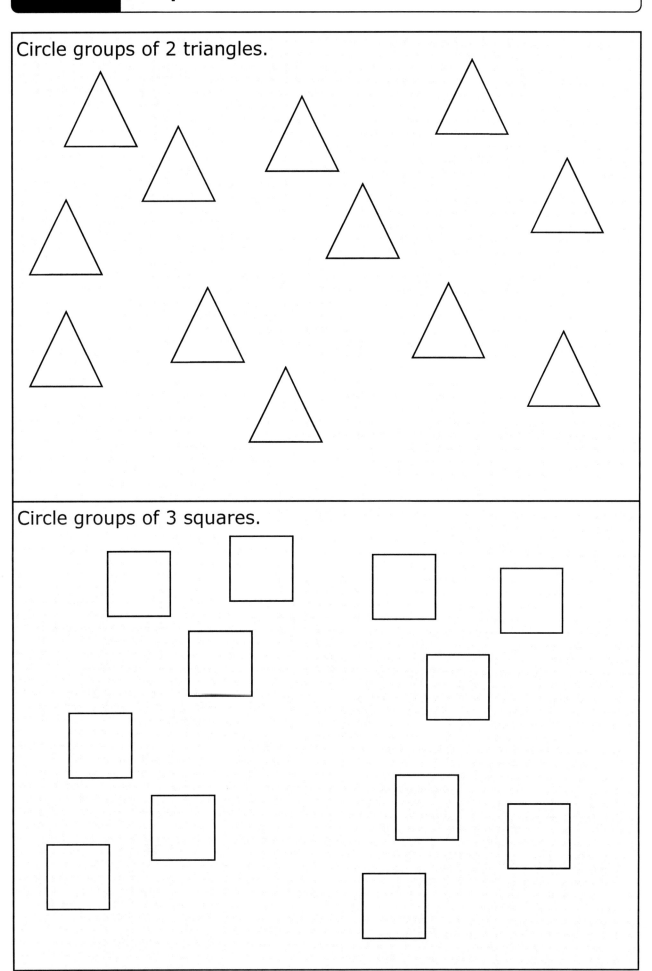

Circle groups of 3 squares.

Colour 1 circle blue.
Colour 3 circles red.

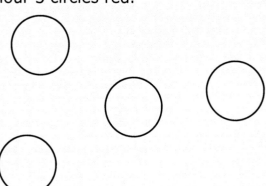

The ratio of blue to red is 1:3.

Colour 2 squares yellow.
Colour 1 square green.

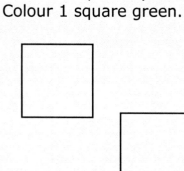

The ratio of yellow to green 2:1.

Colour 5 triangles red.
Colour 2 triangles green.

The ratio of red to green is 5:2.

Colour 2 stars yellow.
Colour 3 stars blue.

The ratio of yellow to blue 2:3.

Colour 1 square purple.
Colour 1 square green.

The ratio of purple to green is 1:1.

Colour 3 pentagons red.
Colour 2 pentagons orange.

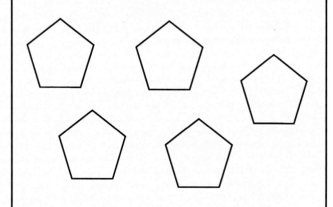

The ratio of red to orange is 3:2.

There are 2 coloured bricks for each white brick.

If there were 6 bricks:

How many would be white? ....

How many bricks would be coloured? ....

If there were 9 bricks:

How many would be white? ....

How many bricks would be coloured? ....

If there were 15 bricks:

How many would be white? ....

How many bricks would be coloured? ....

Draw the 15 bricks.

What is the ratio of stars to triangles? ....

What is the ratio of stars to circles? ....

What is the ratio of stars to squares? ....

What is the ratio of squares to circles? ....

What is the ratio of squares to triangles? ....

What is the ratio of circles to triangles? ....

Colour the correct number of squares in each row.

Colour 1 square = 10% coloured.

Colour 2 square = 20% coloured.

Colour 3 squares = 30% coloured.

Colour 4 squares = 40% coloured.

Colour 5 squares = 50% coloured.

Colour 6 squares = 60% coloured.

Colour 7 squares = 70% coloured.

Colour 8 squares = 80% coloured.

Colour 9 squares = 90% coloured.

Colour 10 squares = 100% coloured.

| N5.6 | Percentages and amounts | 2a |

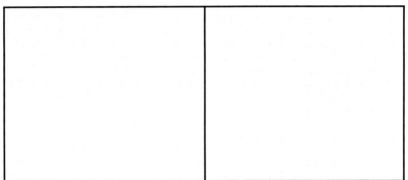

$50\% = \frac{1}{2}$

Here is a bag of 20 marbles.

Share the marbles between Joe and Nina.

Each will have 50% or $\frac{1}{2}$ of 20 marbles.

Joe	Nina

50% of 20 marbles = ... marbles.

$\frac{1}{2}$ of 20 marbles = ... marbles.

$25\% = \frac{1}{4}$

Share the 20 marbles between Joe, Nina, Ajay and Tony.

Each will have 25% or $\frac{1}{4}$

Joe	Nina	Ajay	Tony

25% of 20 marbles = ... marbles.

$\frac{1}{4}$ of 20 marbles = ... marbles.

These squares have 100 tiles. Each tile = 1%.

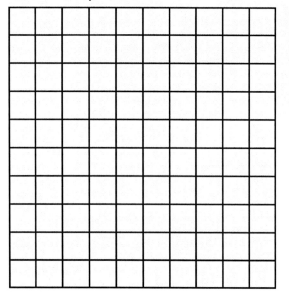

Colour 10% of the tiles.

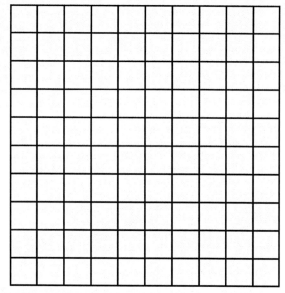

Colour 25% of the tiles.

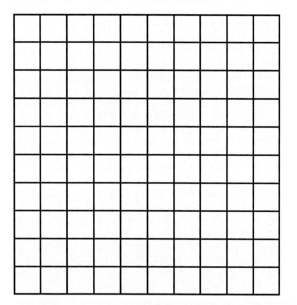

Colour 50% of the tiles.

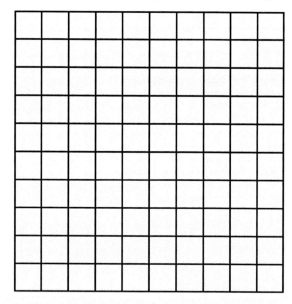

Colour 75% of the tiles.

30 tiles are missing from a floor of 100 tiles. What percentage are missing?	50 out of 100 tiles have been put on the wall. What percentage is this? ....
You need £100 and you have £75. What percentage do you have?	Mum has knitted 80 cm of 100 cm. What percentage has she knitted? ....

D4.1 **Solving data problems** **P6**

Match the stamps.

Find the stamp that matches the stamp in the box on the left.

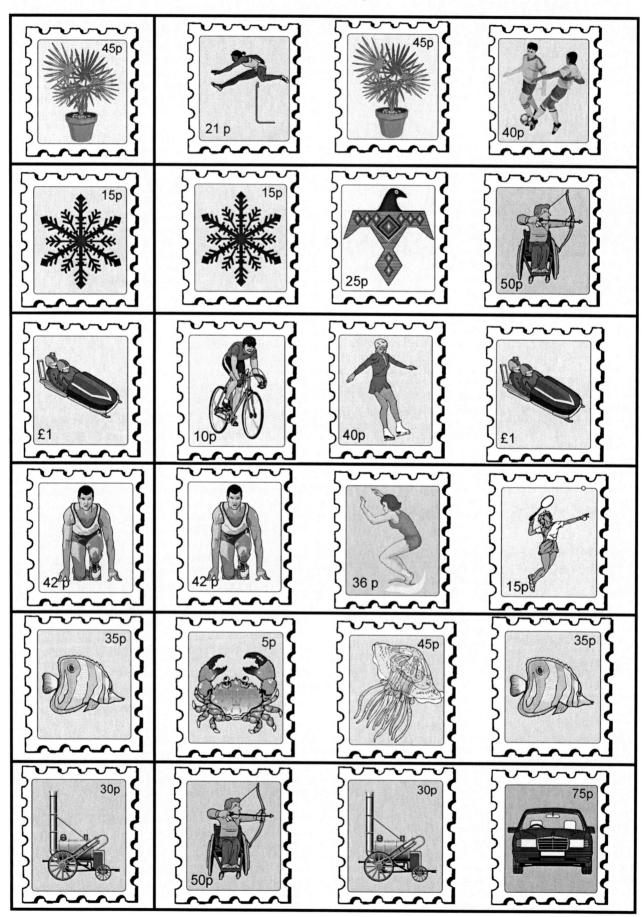

| D4.2 | Grouped data | P7 |

Group the data.

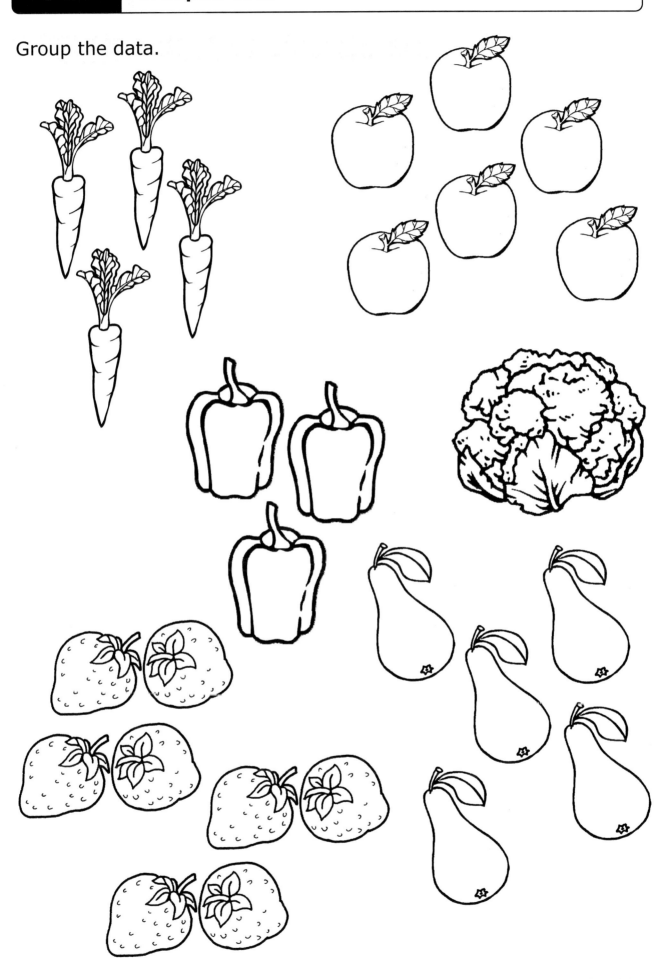

How many in each column? Write below.

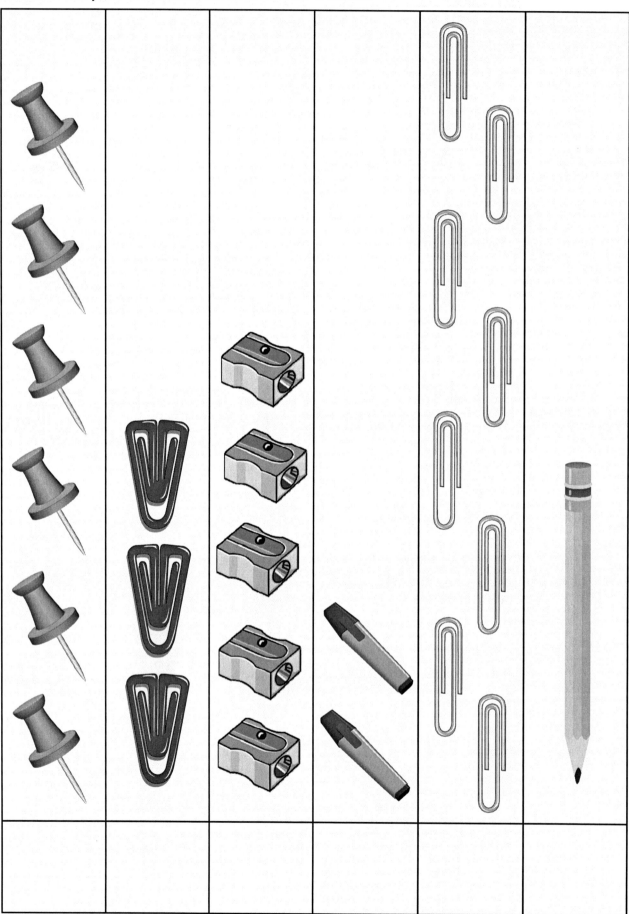

D4.3   **Probability**   **1c**

Tick the correct sentence.

Object	Tick the sentence	Description
		This is the sun.
		This is the moon.
		It is snowing.
		It is windy.
		This is a rainbow.
		This is an umbrella.
		Here is a flower.
		Here is a leaf.
		This is a log.
		This is a tree.
		The tree is covered with leaves
		The tree has no leaves.
		It is dry.
		It is raining.

You had to find the correct answer from .... sentences each time.

Make a choice. Circle the appropriate word to describe the object.

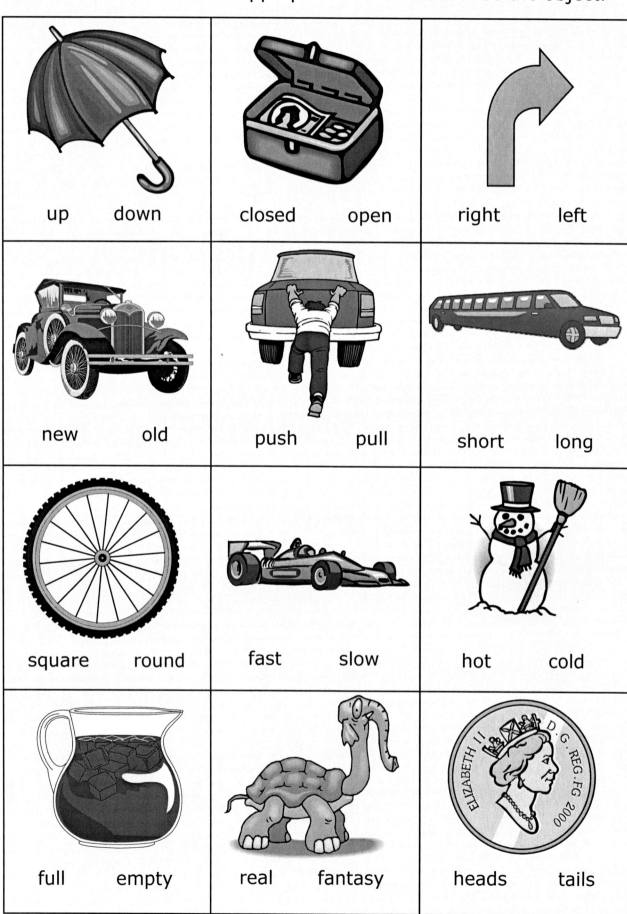

up down	closed open	right left
new old	push pull	short long
square round	fast slow	hot cold
full empty	real fantasy	heads tails

Put a tick or cross by the unlikely pictures.

Read the statement and select the correct probability.

	Impossible	Possible	Certain
It will be dark tomorrow night.			
An elephant will sit in the school hall on Thursday.			
It will rain tomorrow.			
A boat will fly from London to Paris.			
Thursday will follow Wednesday.			
You will eat tomorrow.			
You will have a hamburger tomorrow.			
Your new name will be Enid.			
The sun lights the world.			
Trains run on railway lines.			
Crocodiles sit in trees.			
A book has 150 pages.			

How many gloves in a pair? ....

What is the probability of picking a right hand glove? ....

What is the probability of picking a left hand glove? ....

Half a pack of cards is red and half is black.

What is the probability of picking a red card? ....

What is the probability of picking a black card? ....

Jane has a sweet in one hand.

What is the probability of selecting the correct hand? ....

This is a baby.

How many possibilities? ....

It could be a .... or a ....

Toss a coin.

How many possible outcomes?

What are the possible outcomes?

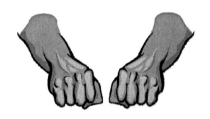

One hand contains 50p and one hand

contains £1.

You select a hand.

How many possible outcomes?

You roll a dice.

How many possible outcomes?

What are the possible outcomes?

There is a green, yellow, red, blue and orange

ball in the bag. Take one ball from the bag.

How many possible outcomes?

What are the possible outcomes?

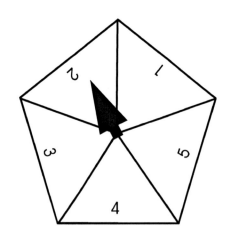

What are the possible outcomes when using a five-sided spinner?

What is the probability of getting an even number?

What is the probability of getting an odd number?

Spin it twice and add the numbers you get. How many ways could you score 8?

What are the possible outcomes when using a six-sided dice?

What is the probability of getting an odd number?

What is the probability of getting an even number?

Roll the dice twice and add the scores. How many ways could you score 7?

Match the numbers. The first one is done.

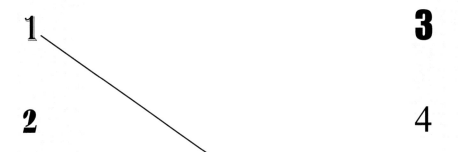

1 — 1

3   4

2   7

4   1

5   2

6   8

7   5

8   9

9   6

Put a string on each balloon.

Give each candle a light.

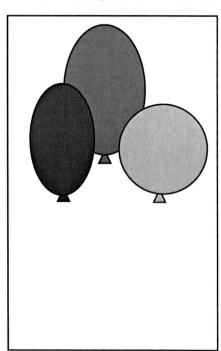

Put 2 candles on each cake.

Put an ice cream in each cone.

Connect each number to its result.

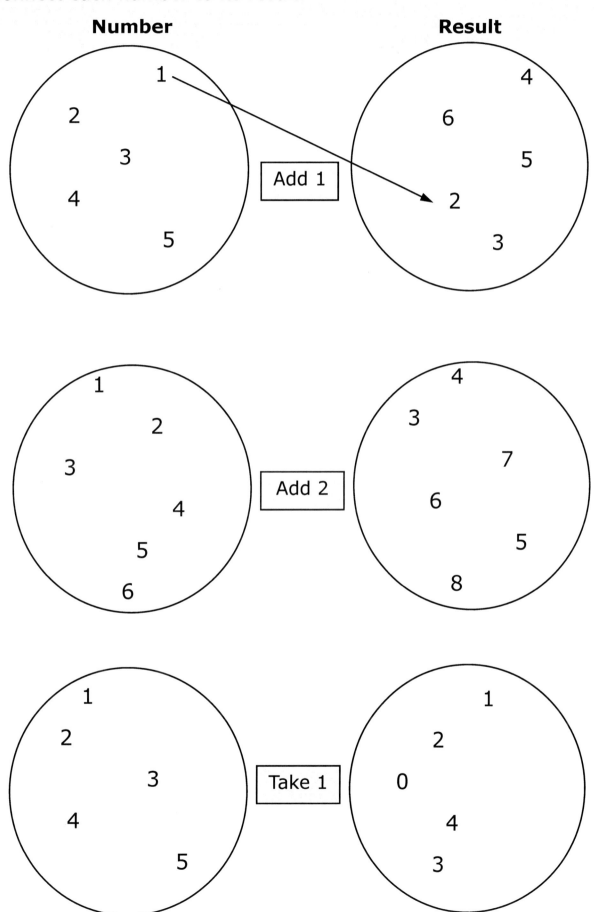

| A5.2 | Coordinates | 1b |

Find the coordinates of these objects.

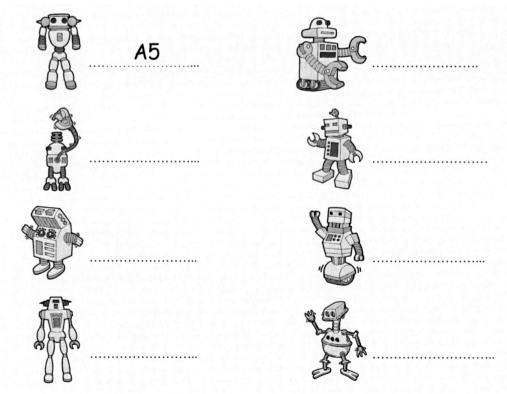

A5.3   **Pairs of values**   **2b**

Complete the values in the tables.

X	1	2	3	4	5	6	7	8	9	10
Y	4	5								

+ 3

X	1	2	3	4	5	6	7	8	9	10
Y										

+ 5

X	1	2	3	4	5	6	7	8	9	10
Y										

+ (6 - 4)

X	1	2	3	4	5	6	7	8	9	10
Y										

2 x

X	1	2	3	4	5	6	7	8	9	10
Y										

5 x

This table represents the 2 times table. The equation is $y = 2x$.
Complete the table, plot the points and draw the graph.

X	1	2	3	4	5	6	7	8	9	10
Y	2	4								

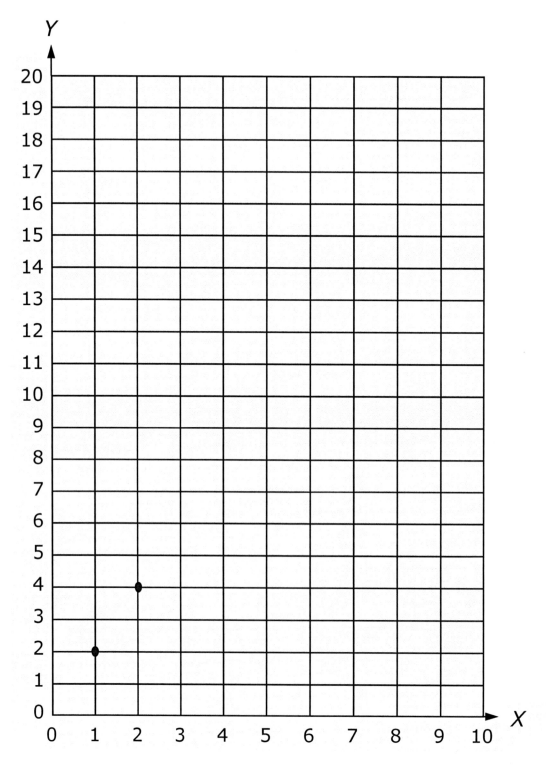

A5.4	**Plotting straight-line graphs**	2a

Plot the number of buttons that are in bags of 2 buttons.

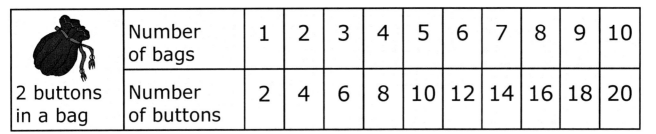

	Number of bags	1	2	3	4	5	6	7	8	9	10
2 buttons in a bag	Number of buttons	2	4	6	8	10	12	14	16	18	20

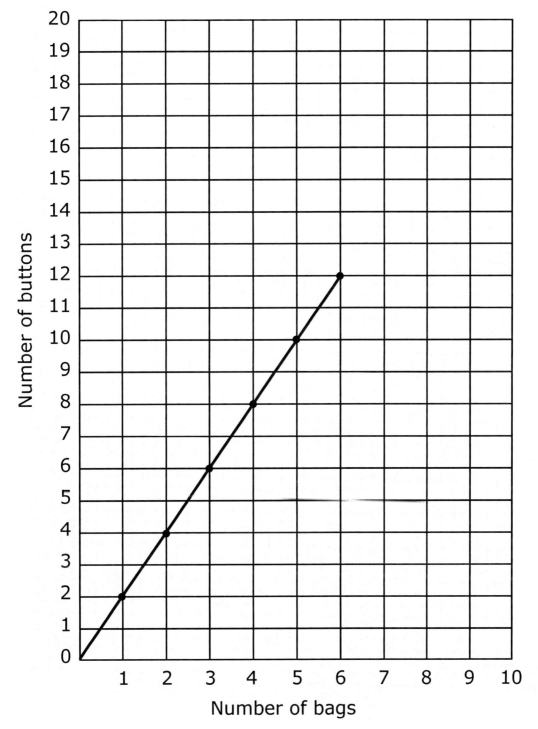

Finish plotting the graph and draw the line.

A5.5 | **Straight line graphs** | **2a**

Complete the tables.

in	35	73	28	84	67	25	79	46
out	40	78						

in	64	32	91	58	16	35	73	27
out								

in	31	54	85	66	37	79	18	23
out								

in	46	38	15	76	49	57	64	85
out								

Fill in the table and plot the graph.
$$Y = X + 2$$

X	1	2	3	4	5
Y					

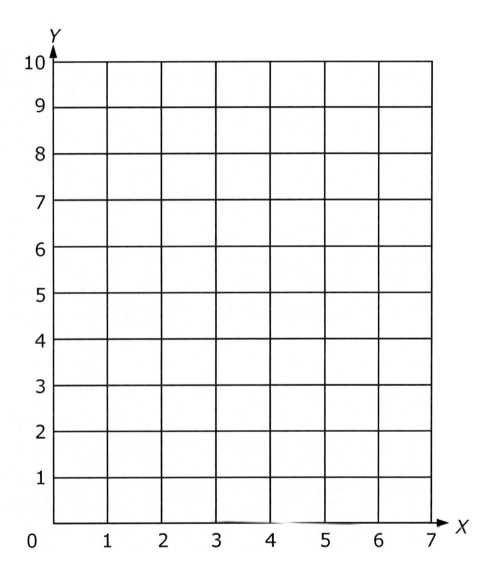

What does X equal when $Y = 10$?

What does Y equal when $X = 6$?

Colour these arrows red.

⇨ Colour these arrows blue.

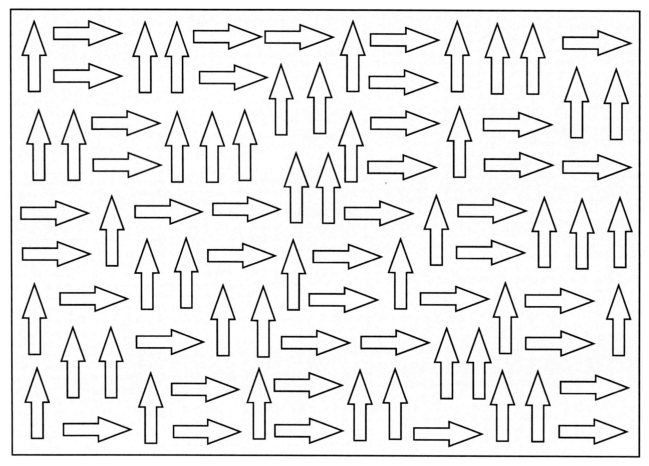

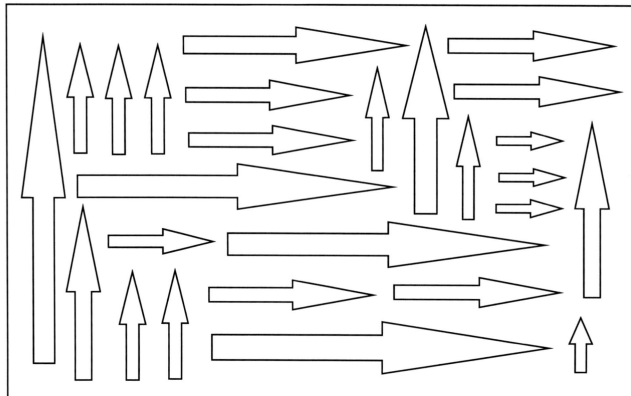

Draw the groups.

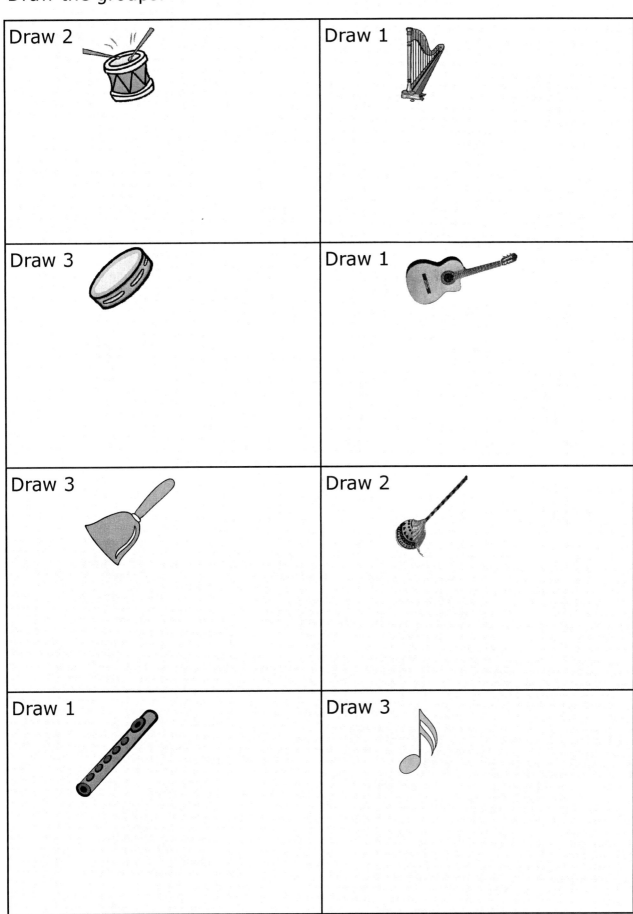

How many of each object?

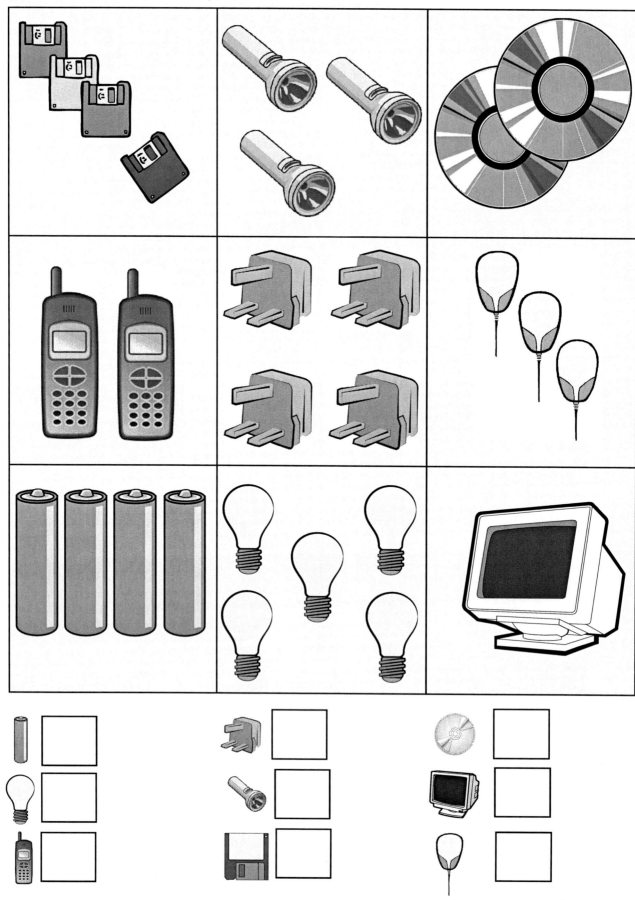

How many in each box? Circle the correct number.

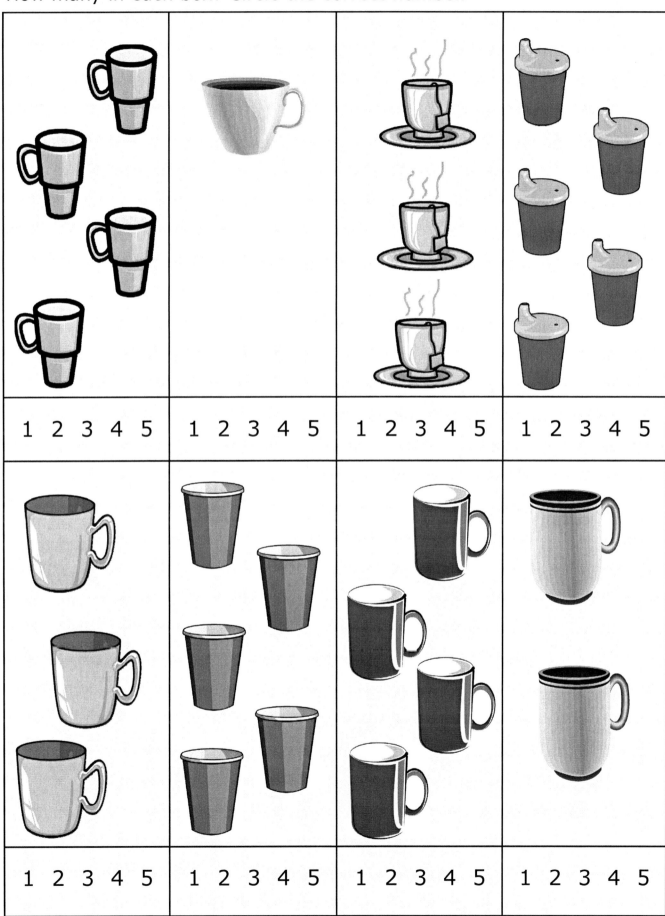

| D5.2 | Calculating statistics | 1c |

Adding on

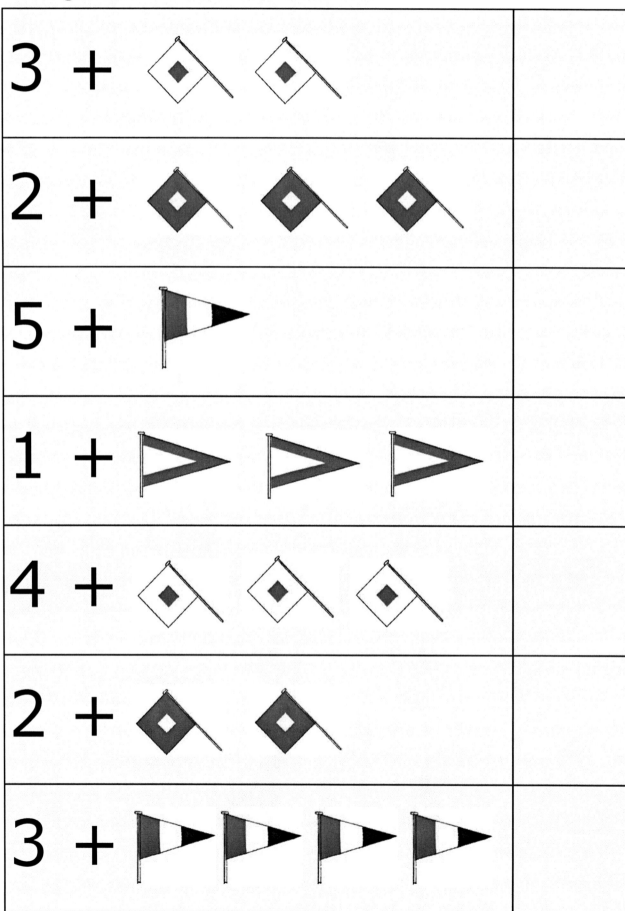

Put the numbers in order starting with the **smallest**.
Circle the number in the **middle**.

4 6 2	3 5 4

2 4 7 3 1	8 2 5 3 6

4 2 1 4 2	5 2 3 5 2

2 4 1 1 2 3 5	6 3 4 6 8 3 4

| D5.3 | Using statistics | 2b |

This shows the number of fast food items
sold between 12 noon and 1 pm.

hamburger									
pizza									
hotdog									
kebab									
sandwich									

Which was the most
popular?

....

Which is the **mode** ?

....

This shows the favourite fruit
of the pupils in a class.

apple	banana	orange	grapes	cherries
6	9	3	7	4

Which is the mode?

....

Which is the least
popular fruit?

....

dog	cat	rabbit	bird	fish

This shows the favourite pet of
pupils in the class

Which is the mode?......................

Which is the least popular pet?

....

This shows the colours of cars parked in the street.

red	blue	silver	black	green	white
3	2	7	5	1	4

Which is the mode? ....

How many cars were parked in the street? ....

This is a weather chart for March 2004

Mon	Tue	Wed	Thu	Fri	Sat	Sun
					1	2
3	4	5	6	7	8	9
10	11	12	13	14	15	16
17	18	19	20	21	22	23
24	25	26	27	28	29	30
31						

Count the number of days for each symbol and fill in the table.

Rain =........ days	Cloudy = ... days
Sun = ... days	Sun/cloud = ... days

What is the mode? ....

Here are the shoe sizes of a group of teenagers.

What is the most common size? ....

What is the mode? ....

37 38 39 40 41 42

Shoe size

D5.4   **Comparing data**   **1a**

Colour the largest number in each pair of numbers.

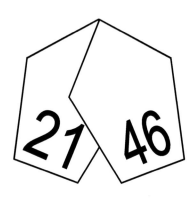

Put the numbers in each box in order starting with the smallest.

17 11 4 13 2	8 12 17 11 6

Put the numbers in each box in order starting with the largest.

12 19 14 7 15	5 14 19 13 10

D5.5	Making conclusions	1a

Number of times the lawn needs cutting each month

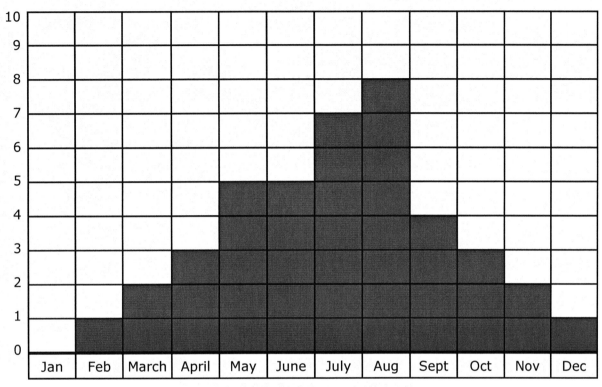

Month

In which month does the lawn need mowing **most** frequently? ....

In which month does the lawn need mowing **least** frequently? ....

Which is the first month of the year when

the grass needs mowing? ....

Which month requires the same number of mowings as April? ....

In which months did the lawn need mowing 5 times? ....

If you need to repair the mower, which month would be best? ....

In which month will there be the most grass cuttings? ....

Why do you think the lawn grows more

in May, June, July and August?

....

If you go on holiday for 2 weeks, what will happen to the lawn?

....

What could you do? ....

D5.6	Discussing findings	2c

The chart shows the colour of the cars in a car park.

	black	white	red	blue	yellow	green	silver	cream	brown

How many cars were in the car park? ....

What was the most common colour? ....

What was the least common colour? ....

There were 3 cars of which colours? ....

There were 2 cars of which colours? ....

There were 5 cars of one colour. Which colour? ....

How many cream cars were there? ....

The most common colour is called the mode.
What is the mode? ....

Were there more green or blue cars? ....

Were there more white or silver cars? ....

How many cars made the mode? ....

How many more silver cars than black cars were there? ....

How many more blue cars than yellow cars were there? ....

G5.1 | **Plans and elevations** | **P7**

Match each object to its shadow.

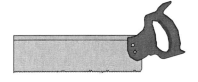

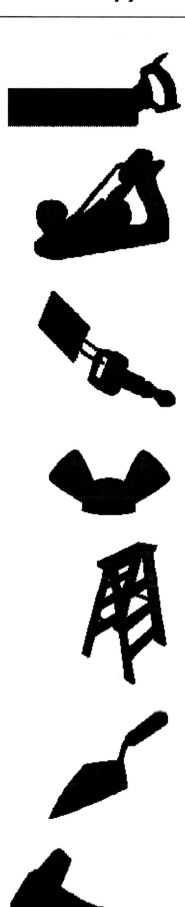

Match.

Match the top view of the object to the side view.

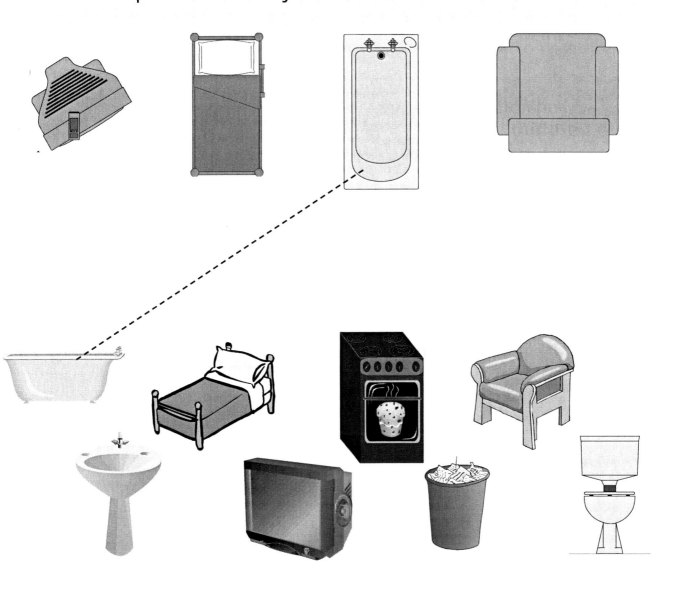

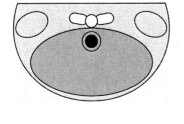

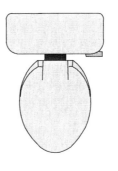

G5.2 **Scale**   **1c**

Draw a **smaller** piece of fruit in each box.

Draw a **larger** vegetable in each box.

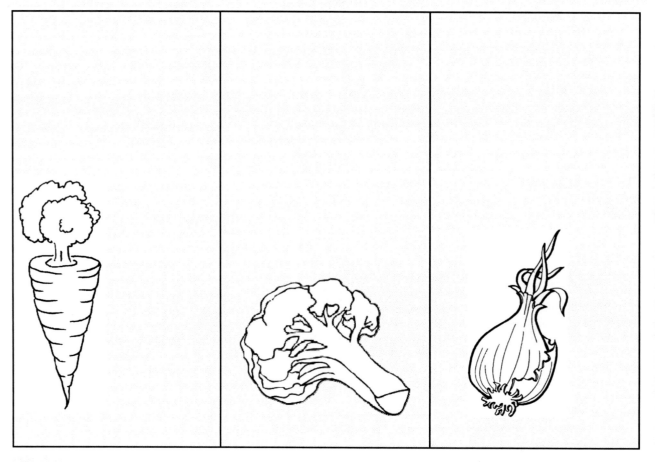

G5.2 | **Scale** | **1b**

How many bricks long is each row of bricks?

   4

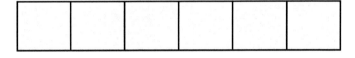

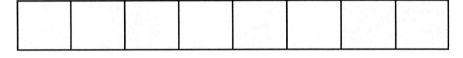

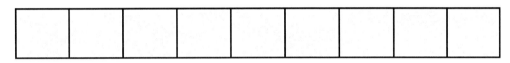

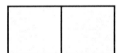

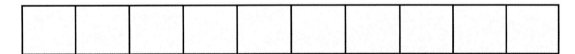

How many bricks long is the screwdriver?

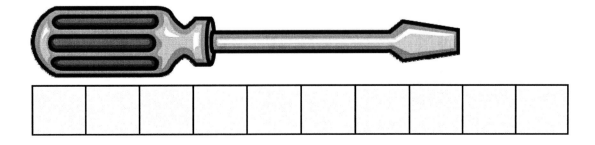

The box

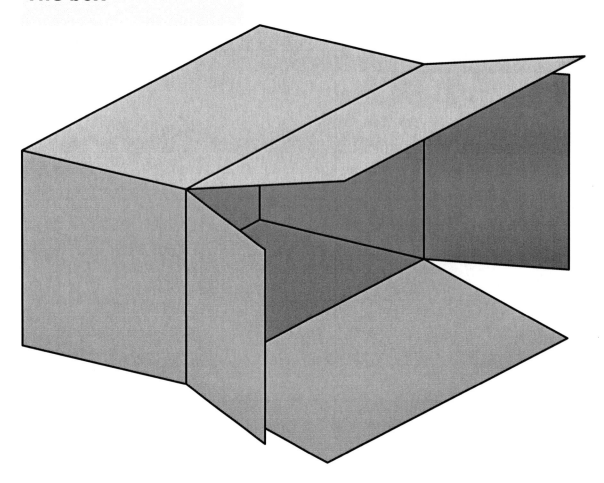

Put a red ball **in** the box.

Put a green ball **under** the box.

Put a blue ball **behind** the box.

Put a yellow ball **beside** the box.

Put a brown brick **on** the box.

Put an orange brick **beside** the box.

Put a red star **above** the box.

Put the objects on the correct shelf at the correct position.

 top shelf
on the **right**

 bottom shelf
on the **right**

 **middle** shelf
on the **left**

 **middle** shelf
in the **middle**

 **bottom** shelf
on the **left**

 top shelf
on the **left**

 middle shelf
on the **right**

 top shelf
in the **middle**

Circle the taller tree.	Circle the smaller leaf.

Circle the wider pot.	Circle the shorter fork.
	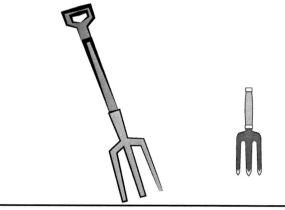
Circle the wider trowel.	Circle the shorter tree.

Circle the smallest flower.	Circle the narrower plant.

| **G5.4** | **Maths Life: A city garden** | **1c** |

The vegetable plot - how many in each row?

	2

How many panels in each metre of fence?

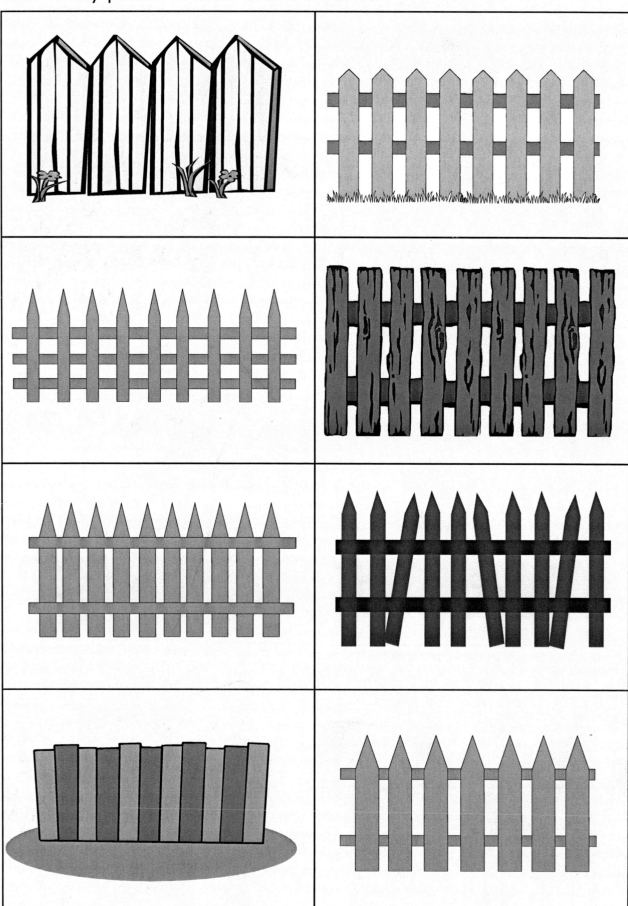